3-D Printing Primer

Contents

Chapter 1

3D printing

For methods of applying a 2D image onto a 3D surface, see pad printing. For methods of copying 2D parallax stereograms that seem 3D to the eye, see lenticular printing and holography.

3D printing, also known as **additive manufacturing (AM)**, refers to various processes used to synthesize a three-dimensional object.[1] In 3D printing, successive layers of material are formed under computer control.[2] These objects can be of almost any shape or geometry, and are produced from a 3D model or other electronic data source. A 3D printer is a type of industrial robot.

3D printing in the term's original sense refers to processes that sequentially deposit material onto a powder bed with inkjet printer heads. More recently, the meaning of the term has expanded to encompass a wider variety of techniques such as extrusion and sintering-based processes. Technical standards generally use the term *additive manufacturing* for this broader sense.

1.1 History

1.1.1 Terminology and methods

Early Additive Manufacturing equipment and materials were developed in the 1980s.[3] In 1981, Hideo Kodama of Nagoya Municipal Industrial Research Institute invented two AM fabricating methods of a three-dimensional plastic model with photo-hardening polymer, where the UV exposure area is controlled by a mask pattern or the scanning fiber transmitter.[4][5] Then in 1984, Chuck Hull of 3D Systems Corporation[6] developed a prototype system based on a process known as stereolithography, in which layers are added by curing photopolymers with ultraviolet light lasers. Hull defined the process as a "system for generating three-dimensional objects by creating a cross-sectional pattern of the object to be formed,"[7][8] but this had been already invented by Kodama. Hull's contribution is the design of the STL (STereoLithography) file format widely accepted by 3D printing software as well as the digital slicing and infill strategies common to many processes today. The term *3D printing* originally referred to a process employing standard and custom inkjet print heads. The technology used by most 3D printers to date—especially hobbyist and consumer-oriented models—is fused deposition modeling, a special application of plastic extrusion.

AM processes for metal sitering or melting (such as selective laser sintering, direct metal laser sintering, and selective laser melting) usually went by their own individual names in the 1980s and 1990s. At the time, nearly all metalworking was produced by casting, fabrication, stamping, and machining; although plenty of automation was applied to those technologies (such as by robot welding and CNC), the idea of a tool or head moving through a 3D work envelope transforming a mass of raw material into a desired shape layer by layer was associated by most people only with processes that removed metal (rather than adding it), such as CNC milling, CNC EDM, and many others. But AM-type sintering was beginning to challenge that assumption. By the mid 1990s, new techniques for material deposition were developed at Stanford and Carnegie Mellon University, including microcasting[9] and sprayed materials.[10] Sacrificial and support materials had also become more common, enabling new object geometries.[11]

The umbrella term *additive manufacturing* gained wider currency in the decade of the 2000s.[12] As the various additive processes matured, it became clear that soon metal removal would no longer be the only metalworking process done under that type of control (a tool or head moving through a 3D work envelope transforming a mass of raw material into a desired shape layer by layer). It was during this decade that the term *subtractive manufacturing* appeared as a retronym for the large family of machining processes with metal removal as their common theme. However, at the time, the term *3D printing* still referred only to the polymer technologies in most minds, and the term *AM* was likelier to be used in metalworking contexts than among polymer/inkjet/stereolithography enthusiasts. The term *subtractive* has not replaced the term *machining*, instead complementing it when a term that covers any removal method is needed.

By the early 2010s, the terms *3D printing* and *additive manufacturing* developed senses in which they were synonymous umbrella terms for all AM technologies. Although this was a departure from their earlier technically narrower senses, it reflects the simple fact that the technologies all share the common theme of sequential-layer material addition/joining throughout a 3D work envelope under automated control. (Other terms that have appeared, which are usually used as AM synonyms (although sometimes as hypernyms), have been *desktop manufacturing*, *rapid manufacturing* [as the logical production-level successor to *rapid prototyping*], and *on-demand manufacturing* [which echoes *on-demand printing* in the 2D sense of *printing*].) The 2010s were the first decade in which metal parts such as engine brackets[13] and large nuts[14] would be grown (either before or instead of machining) in job production rather than obligatorily being machined from bar stock or plate.

With the maturation of the technology, several authors had begun to speculate that 3-D printing could aid in sustainable development in the developing world.[15][16][17]

1.2   General principles

1.2.1   Modeling

Main article: 3D modeling

3D printable models may be created with a computer aided design (CAD) package, via a 3D scanner or by a plain digital camera and photogrammetry software.

The manual modeling process of preparing geometric data for 3D computer graphics is similar to plastic arts such as sculpting. 3D scanning is a process of collecting digital data on the shape and appearance of a real object, creating a digital model based on it.

1.2.2   Printing

Before printing a 3D model from an STL file, it must first be examined for "manifold errors". This step being called the "fixup." Generally STLs that have been produced from a model obtained through 3D scanning often have many manifold errors in them that need to be rectified. Examples of these errors are surfaces that do not connect, or gaps in the models.

Once that is done, the .STL file needs to be processed by a piece of software called a "slicer," which converts the model into a series of thin layers and produces a G-code file containing instructions tailored to a specific type of 3D printer (FDM printers). This G-code file can then be printed with 3D printing client software (which loads the G-code, and uses it to instruct the 3D printer during the 3D printing process).

Printer resolution describes layer thickness and X-Y resolution in dots per inch (dpi) or micrometers (μm). Typical layer thickness is around 100 μm (250 DPI), although some machines can print layers as thin as 16 μm (1,600 DPI).[18] X-Y resolution is comparable to that of laser printers. The particles (3D dots) are around 50 to 100 μm (510 to 250 DPI) in diameter.

Construction of a model with contemporary methods can take anywhere from several hours to several days, depending on the method used and the size and complexity of the model. Additive systems can typically reduce this time to a few hours, although it varies widely depending on the type of machine used and the size and number of models being produced simultaneously.

Traditional techniques like injection moulding can be less expensive for manufacturing polymer products in high quantities, but additive manufacturing can be faster, more flexible and less expensive when producing relatively small quantities of parts. 3D printers give designers and concept development teams the ability to produce parts and concept models using a desktop size printer.

1.2.3 Finishing

Though the printer-produced resolution is sufficient for many applications, printing a slightly oversized version of the desired object in standard resolution and then removing material[19] with a higher-resolution subtractive process can achieve greater precision.

Some printable polymers allow the surface finish to be smoothed and improved using chemical vapor processes.

Some additive manufacturing techniques are capable of using multiple materials in the course of constructing parts. These techniques are able to print in multiple colors and color combinations simultaneously, and would not necessarily require painting.

Some printing techniques require internal supports to be built for overhanging features during construction. These supports must be mechanically removed or dissolved upon completion of the print.

All of the commercialized metal 3-D printers involve cutting the metal component off of the metal substrate after deposition. A new process for the GMAW 3-D printing allows for substrate surface modifications to remove aluminum[20] or steel.[21]

1.3 Processes

Several 3D printing processes have been invented since the late 1970s. The printers were originally large, expensive, and highly limited in what they could produce.[3]

A large number of additive processes are now available. The main differences between processes are in the way layers are deposited to create parts and in the materials that are used. Some methods melt or soften the material to produce the layers, for example. selective laser melting (SLM) or direct metal laser sintering (DMLS), selective laser sintering (SLS), fused deposition modeling (FDM),[22] or fused filament fabrication (FFF), while others cure liquid materials using different sophisticated technologies, such as stereolithography (SLA). With laminated object manufacturing (LOM), thin layers are cut to shape and joined together (e.g. paper, polymer, metal). Each method has its own advantages and drawbacks, which is why some companies offer a choice of powder and polymer for the material used to build the object.[23] Others sometimes use standard, off-the-shelf business paper as the build material to produce a durable prototype. The main considerations in choosing a machine are generally speed, costs of the 3D printer, of the printed prototype, choice and cost of the materials, and color capabilities.[24]

Printers that work directly with metals are generally expensive. However less expensive printers can be used to make a mold, which is then used to make metal parts.[25]

1.3.1 Extrusion deposition

Main article: Fused deposition modeling

Fused deposition modeling (FDM) was developed by S. Scott Crump in the late 1980s and was commercialized in 1990 by Stratasys.[27] After the patent on this technology expired, a large open-source development community developed and both commercial and DIY variants utilizing this type of 3D printer appeared. As a result, the price of this technology has dropped by two orders of magnitude since its creation.

In fused deposition modeling the model or part is produced by extruding small beads of material which harden immediately to form layers. A thermoplastic filament or metal wire that is wound on a coil is unreeled to supply material to an extrusion nozzle head (3D printer extruder). The nozzle head heats the material and turns the flow on and off. Typically stepper motors or servo motors are employed to move the extrusion head and adjust the flow. The printer usually has 3 axes of motion. A computer-aided manufacturing (CAM) software package is used to generate the G-Code that is sent to a microcontroller which controls the motors.

Extrusion in 3-D printing using material extrusion involves a cold end and a hot end.

Various polymers are used, including acrylonitrile butadiene styrene (ABS), polycarbonate (PC), polylactic acid (PLA), high density polyethylene (HDPE), PC/ABS, polyphenylsulfone (PPSU) and high impact polystyrene (HIPS). In general, the polymer is in the form of a filament fabricated from virgin resins. There are multiple projects in the open-sourced community aimed at processing post-consumer plastic waste into filament. These involve machines used to shred and extrude the plastic material into filament.

FDM is somewhat restricted in the variation of shapes that may be fabricated. For example, FDM usually cannot produce stalactite-like structures, since they would be unsupported during the build. Otherwise, a thin support must be designed into the structure which can be broken away during finishing. Fused deposition modeling is also referred to as fused filament fabrication (FFF) by companies who do not hold the original patents like Stratasys does.

1.3.2   Binding of granular materials

Another 3D printing approach is the selective fusing of materials in a granular bed. The technique fuses parts of the layer and then moves downward in the working area, adding another layer of granules and repeating the process until the piece has built up. This process uses the unfused media to support overhangs and thin walls in the part being produced, which reduces the need for temporary auxiliary supports for the piece. A laser is typically used to sinter the media into a solid. Examples include selective laser sintering (SLS), with both metals and polymers (e.g. PA, PA-GF, Rigid GF, PEEK, PS, Alumide, Carbonmide, elastomers), and direct metal laser sintering (DMLS).[28]

Selective Laser Sintering (SLS) was developed and patented by Dr. Carl Deckard and Dr. Joseph Beaman at the University of Texas at Austin in the mid-1980s, under sponsorship of DARPA.[29] A similar process was patented without being commercialized by R. F. Housholder in 1979.[30]

Selective laser melting (SLM) does not use sintering for the fusion of powder granules but will completely melt the powder using a high-energy laser to create fully dense materials in a layer-wise method that has mechanical properties similar to those of conventional manufactured metals.

Electron beam melting (EBM) is a similar type of additive manufacturing technology for metal parts (e.g. titanium alloys). EBM manufactures parts by melting metal powder layer by layer with an electron beam in a high vacuum. Unlike metal sintering techniques that operate below melting point, EBM parts void-freeR.[31][32]

Another method consists of an inkjet 3D printing system. The printer creates the model one layer at a time by spreading a layer of powder (plaster, or resins) and printing a binder in the cross-section of the part using an inkjet-like process. This is repeated until every layer has been printed. This technology allows the printing of full color prototypes, overhangs, and elastomer parts. The strength of bonded powder prints can be enhanced with wax or thermoset polymer impregnation.

1.3.3   Lamination

Main article: Laminated object manufacturing

In some printers, paper can be used as the build material, resulting in a lower cost to print. During the 1990s some companies marketed printers that cut cross sections out of special adhesive coated paper using a carbon dioxide laser and then laminated them together.

In 2005 Mcor Technologies Ltd developed a different process using ordinary sheets of office paper, a tungsten carbide blade to cut the shape, and selective deposition of adhesive and pressure to bond the prototype.[33]

There are also a number of companies selling printers that print laminated objects using thin plastic and metal sheets.

1.3.4 Photopolymerization

Main article: Stereolithography

Stereolithography was patented in 1986 by Chuck Hull.[34] Photopolymerization is primarily used in stereolithography (SLA) to produce a solid part from a liquid. This process was a dramatic departure from the "photosculpture" method of François Willème (1830–1905) developed in 1860 and the photopolymerization of Mitsubishi's Matsubara in 1974.[35]

The "photosculpture" method consisted of photographing a subject from a variety of equidistant angles and projecting each photograph onto a screen, where a pantograph was used to trace the outline onto modeling clay[36][37][38]) In photo-polymerization, a vat of liquid polymer is exposed to controlled lighting under safelight conditions. The exposed liquid polymer hardens. The build plate then moves down in small increments and the liquid polymer is again exposed to light. The process repeats until the model has been built. The liquid polymer is then drained from the vat, leaving the solid model. The EnvisionTEC *Perfactory*[39] is an example of a DLP rapid prototyping system.

Inkjet printer systems like the *Objet PolyJet* system spray photopolymer materials onto a build tray in ultra-thin layers (between 16 and 30 μm) until the part is completed. Each photopolymer layer is cured with UV light after it is jetted, producing fully cured models that can be handled and used immediately, without post-curing. The gel-like support material, which is designed to support complicated geometries, is removed by hand and water jetting. It is also suitable for elastomers.

Ultra-small features can be made with the 3D micro-fabrication technique used in multiphoton photopolymerisation. This approach uses a focused laser to trace the desired 3D object into a block of gel. Due to the nonlinear nature of photo excitation, the gel is cured to a solid only in the places where the laser was focused while the remaining gel is then washed away. Feature sizes of under 100 nm are easily produced, as well as complex structures with moving and interlocked parts.[40]

Yet another approach uses a synthetic resin that is solidified using LEDs.[41]

In Mask-image-projection-based stereolithography a 3D digital model is sliced by a set of horizontal planes. Each slice is converted into a two-dimensional mask image. The mask image is then projected onto a photocurable liquid resin surface and light is projected onto the resin to cure it in the shape of the layer.[42] The technique has been used to create objects composed of multiple materials that cure at different rates.[42] In research systems, the light is projected from below, allowing the resin to be quickly spread into uniform thin layers, reducing production time from hours to minutes.[42] Commercially available devices such as Objet Connex apply the resin via small nozzles.[42]

1.3.5 Metal wire Processes

Laser-based wirefeed systems, such as Laser Metal Deposition-wire (LMD-w), feed wire through a nozzle that is melted by a laser using inert gas shielding in either an open environment (gas surrounding the laser), or in a sealed chamber. Electron beam freeform fabrication uses an electron beam heat source inside a vacuum chamber.

1.4 Printers

1.4.1 Industry use

As October 2012, additive manufacturing systems were on the market that ranged from \$2,000 to \$500,000 in price and were employed in industries including aerospace, architecture, automotive, defense, and medical replacements, among many others. For example, General Electric uses the high-end model to build parts for turbines.[43] Many of these systems are used for rapid prototyping, before mass production methods are employed.

1.4.2   Consumer use

Several projects and companies are making efforts to develop affordable 3D printers for home desktop use. Much of this work has been driven by and targeted at DIY/Maker/enthusiast/early adopter communities, with additional ties to the academic and hacker communities.[44]

RepRap Project is one of the longest running projects in the desktop category. The RepRap project aims to produce a free and open source hardware (FOSH) 3D printer, whose full specifications are released under the GNU General Public License, and which is capable of replicating itself by printing many of its own (plastic) parts to create more machines.[45][46] RepRaps have already been shown to be able to print circuit boards[47] and metal parts.[48][49]

Because of the FOSH aims of RepRap, many related projects have used their design for inspiration, creating an ecosystem of related or derivative 3D printers, most of which are also open source designs. The availability of these open source designs means that variants of 3D printers are easy to invent. The quality and complexity of printer designs, however, as well as the quality of kit or finished products, varies greatly from project to project. This rapid development of open source 3D printers is gaining interest in many spheres as it enables hyper-customization and the use of public domain designs to fabricate open source appropriate technology. This technology can also assist initiatives in sustainable development since technologies are easily and economically made from resources available to local communities.[15][16]

The cost of 3D printers has decreased dramatically since about 2010, with machines that used to cost $20,000 now costing less than $1,000.[50] For instance, as of 2013, several companies and individuals are selling parts to build various RepRap designs, with prices starting at about €400 / US$500.[51] The open source Fab@Home project[52] has developed printers for general use with anything that can be squirted through a nozzle, from chocolate to silicone sealant and chemical reactants. Printers following the project's designs have been available from suppliers in kits or in pre-assembled form since 2012 at prices in the US$2000 range.[51] The Kickstarter funded Peachy Printer is designed to cost $100[53] and several other new 3D printers are aimed at the small, inexpensive market including the mUVe3D and Lumifold. Rapide 3D has designed a professional grade crowdsourced 3D-printer costing $1499 which has no fumes nor constant rattle during use.[54] The 3Doodler, "3D printing pen", raised $2.3 million on Kickstarter with the pens selling at $99,[55] though the 3D Doodler has been criticized for being more of a crafting pen than a 3D printer.[56]

As the costs of 3D printers have come down they are becoming more appealing financially to use for self-manufacturing of personal products.[57] In addition, 3D printing products at home may reduce the environmental impacts of manufacturing by reducing material use and distribution impacts.[58]

In addition, several RecycleBots such as the commercialized Filastruder have been designed and fabricated to convert waste plastic, such as shampoo containers and milk jugs, into inexpensive RepRap filament.[59] There is some evidence that using this approach of distributed recycling is better for the environment.[60]

The development and hyper-customization of the RepRap-based 3D printers has produced a new category of printers suitable for small business and consumer use. Manufacturers such as Solidoodle,[43] Robo 3D, RepRapPro and Pirx 3D have introduced models and kits priced at less than $1,000, thousands less than they were in September 2012.[43] Depending on the application, the print resolution and speed of manufacturing lies somewhere between a personal printer and an industrial printer. A list of printers with pricing and other information is maintained.[51] Most recently delta robots, like the TripodMaker, have been utilized for 3D printing to increase fabrication speed further.[61] For delta 3D printers, due to its geometry and differentiation movements, the accuracy of the print depends on the position of the printer head.[62]

Some companies are also offering software for 3D printing, as a support for hardware manufactured by other companies.[63]

1.4.3   Large 3D printers

Large 3D printers have been developed for industrial, education, and demonstrative uses. A large delta-style 3D printer was built in 2014 by SeeMeCNC. The printer is capable of making an object with diameter of up to 4 feet (1.2 m) and up to 10 feet (3.0 m) in height. It also uses plastic pellets as the raw material instead of the typical plastic filaments used in other 3D printers.[64]

Another type of large printer is Big Area Additive Manufacturing (BAAM). The goal is to develop printers that can produce a large object in high speed. A BAAM machine of Cincinnati Incorporated can produce an object at the speeds 200-500 times faster than typical 3D printers available in 2014. Another BAAM machine is being developed by Lockheed

Martin with an aim to print long objects of up to 100 feet (30 m) to be used in aerospace industries.[65]

1.4.4 Microscale and nanoscale 3D printing

Microelectronic device fabrication methods can be employed to perform the 3D printing of nanoscale-size objects. Such printed objects are typically grown on a solid substrate, e.g. silicon wafer, to which they adhere after printing as they're too small and fragile to be manipulated post-construction.

In one technique, 3D nanostructures can be printed by physically moving a dynamic stencil mask during the material deposition process, somewhat analogous to the extrusion method of traditional 3D printers. Programmable-height nanostructures with resolutions as small as 10 nm have been produced in this fashion, by metallic physical vapor deposition through a piezo-actuator controlled stencil mask having a milled nanopore in a silicon nitride membrane.[66]

Another method enhances the photopolymerization process on a much smaller scale, using finely-focused lasers controlled by adjustable mirrors. This method has produced objects with feature resolutions of 100 nm.[67]

1.5 Manufacturing applications

> Three-dimensional printing makes it as cheap to create single items as it is to produce thousands and thus undermines economies of scale. It may have as profound an impact on the world as the coming of the factory did....Just as nobody could have predicted the impact of the steam engine in 1750—or the printing press in 1450, or the transistor in 1950—it is impossible to foresee the long-term impact of 3D printing. But the technology is coming, and it is likely to disrupt every field it touches.
> — *The Economist*, in a February 10, 2011 leader[68]

AM technologies found applications starting in the 1980s in product development, data visualization, rapid prototyping, and specialized manufacturing. Their expansion into production (job production, mass production, and distributed manufacturing) has been under development in the decades since. Industrial production roles within the metalworking industries[69] achieved significant scale for the first time in the early 2010s. Since the start of the 21st century there has been a large growth in the sales of AM machines, and their price has dropped substantially.[70] According to Wohlers Associates, a consultancy, the market for 3D printers and services was worth $2.2 billion worldwide in 2012, up 29% from 2011.[71] There are many applications for AM technologies, including architecture, construction (AEC), industrial design, automotive, aerospace,[72] military, engineering, dental and medical industries, biotech (human tissue replacement), fashion, footwear, jewelry, eyewear, education, geographic information systems, food, and many other fields.

Additive manufacturing's earliest applications have been on the toolroom end of the manufacturing spectrum. For example, rapid prototyping was one of the earliest additive variants, and its mission was to reduce the lead time and cost of developing prototypes of new parts and devices, which was earlier only done with subtractive toolroom methods such as cnc milling and turning, and precision grinding, far more accurate than 3d printing with accuracy down to 0.00005" and creating better quality parts faster, but sometimes too expensive for low accuracy prototype parts.[73] With technological advances in additive manufacturing, however, and the dissemination of those advances into the business world, additive methods are moving ever further into the production end of manufacturing in creative and sometimes unexpected ways.[73] Parts that were formerly the sole province of subtractive methods can now in some cases be made more profitably via additive ones. In addition, new developments in RepRap technology allow the same device to perform both additive and subtractive manufacturing by swapping magnetic-mounted tool heads.[74]

1.5.1 Distributed manufacturing

Main article: 3D printing marketplace

Additive manufacturing in combination with cloud computing technologies allows decentralized and geographically independent distributed production.[75] Distributed manufacturing as such is carried out by some enterprises; there is also a services like 3D Hubs that put people needing 3D printing in contact with owners of printers.[76]

Some companies offer on-line 3D printing services to both commercial and private customers,[77] working from 3D designs uploaded to the company website. 3D-printed designs are either shipped to the customer or picked up from the service provider.[78]

1.5.2   Mass customization

Companies have created services where consumers can customize objects using simplified web based customisation software, and order the resulting items as 3D printed unique objects.[79][80] This now allows consumers to create custom cases for their mobile phones.[81] Nokia has released the 3D designs for its case so that owners can customize their own case and have it 3D printed.[82]

1.5.3   Rapid manufacturing

Advances in RP technology have introduced materials that are appropriate for final manufacture, which has in turn introduced the possibility of directly manufacturing finished components. One advantage of 3D printing for rapid manufacturing lies in the relatively inexpensive production of small numbers of parts.

Rapid manufacturing is a new method of manufacturing and many of its processes remain unproven. 3D printing is now entering the field of rapid manufacturing and was identified as a "next level" technology by many experts in a 2009 report.[83] One of the most promising processes looks to be the adaptation of selective laser sintering (SLS), or direct metal laser sintering (DMLS) some of the better-established rapid prototyping methods. As of 2006, however, these techniques were still very much in their infancy, with many obstacles to be overcome before RM could be considered a realistic manufacturing method.[84]

1.5.4   Rapid prototyping

Main article: rapid prototyping

Industrial 3D printers have existed since the early 1980s and have been used extensively for rapid prototyping and research purposes. These are generally larger machines that use proprietary powdered metals, casting media (e.g. sand), plastics, paper or cartridges, and are used for rapid prototyping by universities and commercial companies.

1.5.5   Research

3D printing can be particularly useful in research labs due to its ability to make specialized, bespoke geometries. In 2012 a proof of principle project at the University of Glasgow, UK, showed that it is possible to use 3D printing techniques to assist in the production of chemical compounds. They first printed chemical reaction vessels, then used the printer to deposit reactants into them.[85] They have produced new compounds to verify the validity of the process, but have not pursued anything with a particular application.[85]

1.5.6   Food

Cornell Creative Machines Lab announced in 2012 that it was possible to produce customized food with 3D Hydrocolloid Printing.[86] Additive manufacturing of food is currently being developed by squeezing out food, layer by layer, into three-dimensional objects. A large variety of foods are appropriate candidates, such as chocolate and candy, and flat foods such as crackers, pasta,[87] and pizza.[88] NASA has considered the versatility of the concept, awarding a contract to the Systems and Materials Research Consultancy to study the feasibility of printing food in space.[89]

1.5.7   Medical Applications

Professor Leroy Cronin of Glasgow University proposed in a 2012 TED Talk that it was possible to use chemical inks to print medicine.[90] Similarly, 3D printing has been considered as a method of implanting stem cells capable of generating new tissues and organs in living humans.[91]

1.6   Industrial applications

1.6.1   Apparel

3D printing has spread into the world of clothing with fashion designers experimenting with 3D-printed bikinis, shoes, and dresses.[92] In commercial production Nike is using 3D printing to prototype and manufacture the 2012 Vapor Laser Talon football shoe for players of American football, and New Balance is 3D manufacturing custom-fit shoes for athletes.[92][93]

3D printing has come to the point where companies are printing consumer grade eyewear with on-demand custom fit and styling (although they cannot print the lenses). On-demand customization of glasses is possible with rapid prototyping.[94]

1.6.2   Vehicle

In early 2014, the Swedish supercar manufacturer, Koenigsegg, announced the One:1, a supercar that utilizes many components that were 3D printed. In the limited run of vehicles Koenigsegg produces, the One:1 has side-mirror internals, air ducts, titanium exhaust components, and even complete turbocharger assemblies that have been 3D printed as part of the manufacturing process.[95]

Urbee is the name of the first car in the world car mounted using the technology 3D printing (his bodywork and his car windows were "printed"). Created in 2010 through the partnership between the US engineering group Kor Ecologic and the company Stratasys (manufacturer of printers Stratasys 3D), it is a hybrid vehicle with futuristic look.[96][97][98]

In May 2015 Airbus announced that its new Airbus A350 XWB included over 1000 components manufactured by 3D printing.

3D printing is also being utilized by air forces to print spare parts for planes. In 2015, a Royal Air Force Eurofighter Typhoon fighter jet flew with printed parts. The United States Air Force has begun to work with 3D printers, and the Israeli Air Force has also purchased a 3D printer to print spare parts.[99]

1.6.3   Construction

Main article: Building printing

Until recent years models were built by hand, often taking a long time. Thus, architects are often forced to show their clients drawings of their projects. According to Erik Kinipper, clients usually need to see the product from all possible viewpoints in space to get a clearer picture of the design and make an informed decision. In order to get these scale models to clients in a small amount of time, architects and architecture firms tend to rely on 3D printing.[100] Using 3D printing, these firms can reduce lead times of production by 50 to 80 percent, producing scale models up to 60 percent lighter than the machined part while being sturdy.[101] Thus, the designs and the models are only limited by a person's imagination.

The improvements on accuracy, speed and quality of materials in 3D printing technology have opened new doors for it to move beyond the use of 3-D printing in the modeling process and actually move it to manufacturing strategy. A good example is Dr. Behrokh Khoshnevis' research at the University of Southern California which resulted in a 3D printer that can build a house in 24 hours .The process is called Contour Crafting. Khoshnevis, Russell, Kwon, & Bukkapatnam, define contour crafting as an additive manufacturing process which uses computer controlled systems to repeatedly lay down layers of materials such as concrete. Bushey also discussed Khoshnevis's robot which comes equipped with a nozzle that spews out concrete and can build a home based on a set computer pattern. Contour Crafting technology has great potential for automating the construction of whole structures as well as sub-components. Using this process, a single house or a colony of houses, each with possibly a different design, may be automatically constructed in a single run, embedded in each house all the conduits for electrical, plumbing and air-conditioning.[102]

1.6.4   Firearms

Main article: 3D printed firearms

In 2012, the US-based group Defense Distributed disclosed plans to "[design] a working plastic gun that could be downloaded and reproduced by anybody with a 3D printer."[103][104] Defense Distributed has also designed a 3D printable AR-15 type rifle lower receiver (capable of lasting more than 650 rounds) and a 30 round M16 magazine The AR-15 has multiple receivers (both an upper and lower receiver), but the legally controlled part is the one that is serialized (the lower, in the AR-15's case). Soon after Defense Distributed succeeded in designing the first working blueprint to produce a plastic gun with a 3D printer in May 2013, the United States Department of State demanded that they remove the instructions from their website.[105] After Defense Distributed released their plans, questions were raised regarding the effects that 3D printing and widespread consumer-level CNC machining[106][107] may have on gun control effectiveness.[108][109][110][111]

In 2014, a man from Japan became the first person in the world to be imprisoned for making 3D printed firearms.[112] Yoshitomo Imura posted videos and blueprints of the gun online and was sentenced to jail for two years. Police found at least two guns in his household that were capable of firing bullets.[112]

1.6.5   Medical

3D printing has been used to print patient specific implant and device for medical use. Successful operations include a titanium pelvis implanted into a British patient, titanium lower jaw transplanted to a Belgian patient,[113] and a plastic tracheal splint for an American infant.[114] The hearing aid and dental industries are expected to be the biggest area of future development using the custom 3D printing technology.[115] In March 2014, surgeons in Swansea used 3D printed parts to rebuild the face of a motorcyclist who had been seriously injured in a road accident.[116] Research is also being conducted on methods to bio-print replacements for lost tissue due to arthritis and cancer.

The things we are printing are becoming more personal and intimate overtime. This is especially true even in medicine: increasingly, what we are printing is ourselves. 3D printing is a great technological advancement in the medical field. 3D printing technology can now be used to make exact replicas of organs. Doctors can use these to plan out surgeries a lot more accurately. While, medical students may even be able to perform practice operations. The printer uses images from patients' MRI or CT scan images as a template and lays down layers of rubber or plastic. These types of procedures are becoming more and more common among doctors and medical researchers everywhere. Many have begun to use 3D printing in order to accurately undergo surgical procedures. This helps that may have been considered inoperable before have the opportunity to have a successful surgery. The hope is that these printers will eventually be ableto produce actual organs that can be transplanted into patients by replacing the rubber and plastic printer "ink" with human cells. "The printer and software usually cost in the range of $100,000." Which happens to be less than a CT scan or MRI setup. Rader said "He predicts that interest in the technology will continue to grow as research shows how using simulated organs leads to better surgical outcomes and shorter operating times." This will also help the patients feel at ease as they can be provided with a live model to better explain surgical procedures and put the patient at ease.[117]

Medical devices

In October 24, 2014, a five-year-old girl born without fully formed fingers on her left hand became the first child in the UK to have a prosthetic hand made with 3D printing technology. Her hand was designed by US-based E-nable, an open source design organisation which uses a network of volunteers to design and make prosthetics mainly for children. The prosthetic hand was based on a plaster cast made by her parents.[118] A boy named Alex was also born with a missing arm from just above the elbow. The team was able to use 3D printing to upload an e-NABLE Myoelectric arm that runs off of servos and batteries that are actuated by the electromyography muscle. With the use of 3D printers, E-NABLE has so far distributed more than 400 plastic hands to children.

Printed prosthetics have been used in rehabilitation of crippled animals. In 2013, a 3D printed foot let a crippled duckling walk again.[119] In 2014 a chihuahua born without front legs was fitted with a harness and wheels created with a 3D printer.[120] 3D printed hermit crab shells let hermit crabs inhabit a new style home.[121] A prosthetic beak was another tool developed by the use of 3D printing to help aid a bald eagle named Beauty, whose beak was severely mutilated from a shot in the face. Since 2014, commercially available titanium knee implants made with 3D printer for dogs have been used to restore the animal mobility. Over 10,000 dogs in Europe and United States have been treated after only one year.[122]

In February 2015, FDA approved the marketing of a surgical bolt which facilitates less-invasive foot surgery and eliminates the need to drill through bone. The 3-D printed titanium device, 'FastForward Bone Tether Plate' is approved to use in correction surgery to treat bunion.[123]

Bio-printing

As of 2012, 3D bio-printing technology has been studied by biotechnology firms and academia for possible use in tissue engineering applications in which organs and body parts are built using inkjet techniques. In this process, layers of living cells are deposited onto a gel medium or sugar matrix and slowly built up to form three-dimensional structures including vascular systems.[124] The first production system for 3D tissue printing was delivered in 2009, based on NovoGen bio-printing technology.[125] Several terms have been used to refer to this field of research: organ printing, bio-printing, body part printing,[126] and computer-aided tissue engineering, among others.[127] The possibility of using 3D tissue printing to create soft tissue architectures for reconstructive surgery is also being explored.[128]

In 2013, Chinese scientists began printing ears, livers and kidneys, with living tissue. Researchers in China have been able to successfully print human organs using specialized 3D bio printers that use living cells instead of plastic. Researchers at Hangzhou Dianzi University designed the "3D bio printer" dubbed the "Regenovo". Xu Mingen, Regenovo's developer, said that it takes the printer under an hour to produce either a mini liver sample or a four to five inch ear cartilage sample. Xu also predicted that fully functional printed organs may be possible within the next ten to twenty years.[129][130] In the same year, researchers at the University of Hasselt, in Belgium had successfully printed a new jawbone for an 83-year-old Belgian woman.[131]

Pills

The first pill manufactured by 3D printing was approved by FDA in August 2015. Binder-jetting into a powder bed of the drug allows very porous pills to be produced, which enables high drug doses in a single pill which dissolves quickly and can be ingested easily.[132] This has been demonstrated for Spritam, a reformulation of levetiracetam for the treatment of epilepsy.

1.6.6 Computers and robots

See also: Modular design and Open-source robotics

3D printing can be used to make laptops and other computers, including cases, as Novena and VIA OpenBook standard laptop cases. I.e. a Novena motherboard can be bought and be used in a printed VIA OpenBook case.[133]

Open-source robots are built using 3D printers. Double Robotics grant access to their technology (an open SDK).[134][135][136] On the other hand, 3&DBot is an Arduino 3D printer-robot with wheels[137] and ODOI is a 3D printed humanoid robot.[138]

1.6.7 Space

See also: 3D-printed spacecraft and 3D printing § Construction

In September 2014, SpaceX delivered the first zero-gravity 3-D printer to the International Space Station (ISS). On December 19, 2014, NASA emailed CAD drawings for a socket wrench to astronauts aboard the ISS, who then printed the tool using its 3-D printer. Applications for space offer the ability to print parts or tools on-site, as opposed to using rockets to bring along pre-manufactured items for space missions to human colonies on the moon, Mars, or elsewhere.[139] The European Space Agency plans to deliver its new Portable On-Board 3D Printer (POP3D for short) to the International Space Station by June 2015, making it the second 3D printer in space.[140][141]

Furthermore, the Sinterhab project is researching a lunar base constructed by 3D printing using lunar regolith as a base material. Instead of adding a binding agent to the regolith, researchers are experimenting with microwave sintering to create solid blocks from the raw material.[142]

Similar researches and projects like these could allow faster construction for lower costs, and has been investigated for construction of off-Earth habitats.[143][144]

1.7 Sociocultural applications

In 2005, a rapidly expanding hobbyist and home-use market was established with the inauguration of the open-source RepRap and Fab@Home projects. Virtually all home-use 3D printers released to-date have their technical roots in the ongoing RepRap Project and associated open-source software initiatives.[145] In distributed manufacturing, one study has found[146] that 3D printing could become a mass market product enabling consumers to save money associated with purchasing common household objects.[57] For example, instead of going to a store to buy an object made in a factory by injection molding (such as a measuring cup or a funnel), a person might instead print it at home from a downloaded 3D model.

1.7.1 Art

In 2005, academic journals had begun to report on the possible artistic applications of 3D printing technology.[147] By 2007 the mass media followed with an article in the Wall Street Journal[148] and Time Magazine, listing a 3D printed design among their 100 most influential designs of the year.[149] During the 2011 London Design Festival, an installation, curated by Murray Moss and focused on 3D Printing, was held in the Victoria and Albert Museum (the V&A). The installation was called *Industrial Revolution 2.0: How the Material World will Newly Materialize*.[150]

Some of the recent developments in 3D printing were revealed at the 3DPrintshow in London, which took place in November 2013 and 2014. The art section had in exposition artworks made with 3D printed plastic and metal. Several artists such as Joshua Harker, Davide Prete, Sophie Kahn, Helena Lukasova, Foteini Setaki showed how 3D printing can modify aesthetic and art processes. One part of the show focused on ways in which 3D printing can advance the medical field. The underlying theme of these advances was that these printers can be used to create parts that are printed with specifications to meet each individual. This makes the process safer and more efficient. One of these advances is the use of 3D printers to produce casts that are created to mimic the bones that they are supporting. These custom-fitted casts are open, which allow the wearer to scratch any itches and also wash the damaged area. Being open also allows for open ventilation. One of the best features is that they can be recycled to create more casts.[151]

3D printing is becoming more popular in the customisable gifts industry, with products such as personalized mobile phone cases and dolls,[152] as well as 3D printed chocolate.[153]

The use of 3D scanning technologies allows the replication of real objects without the use of moulding techniques that in many cases can be more expensive, more difficult, or too invasive to be performed, particularly for precious or delicate cultural heritage artifacts[154] where direct contact with the moulding substances could harm the original object's surface.

Critical making refers to the hands on productive activities that link digital technologies to society. It is invented to bridge the gap between creative physical and conceptual exploration.[155] The term was popularized by Matt Ratto, an Assistant Professor and director of the Critical Making lab in the Faculty of Information at the University of Toronto. Ratto describes one of the main goals of critical as "to use material forms of engagement with technologies to supplement and extend critical reflection and, in doing so, to reconnect our lived experiences with technologies to social and conceptual critique".[156] The main focus of critical making is open design,[157] which includes, in addition to 3D printing technologies, also other digital software and hardware. People usually reference spectacular design when explaining critical making.[158]

1.7.2 Communication

Employing additive layer technology offered by 3D printing, Terahertz devices which act as waveguides, couplers and bends have been created. The complex shape of these devices could not be achieved using conventional fabrication techniques. Commercially available professional grade printer EDEN 260V was used to create structures with minimum feature size of 100 μm. The printed structures were later DC sputter coated with gold (or any other metal) to create a Terahertz Plasmonic Device.[159]

1.7.3 Domestic use

As of 2012, domestic 3D printing was mainly practiced by hobbyists and enthusiasts, and was little used for practical household applications. A working clock was made,[160] and gears were printed for home woodworking machines among other purposes.[161] 3D printing was also used for ornamental objects. Web sites associated with home 3D printing tended to include backscratchers, coathooks, doorknobs etc.[162]

The open source Fab@Home project[52] has developed printers for general use. They have been used in research environments to produce chemical compounds with 3D printing technology, including new ones, initially without immediate application as proof of principle.[85] The printer can print with anything that can be dispensed from a syringe as liquid or paste. The developers of the chemical application envisage both industrial and domestic use for this technology, including enabling users in remote locations to be able to produce their own medicine or household chemicals.[163][164]

3D printing is now working its way into households, and more and more children are being introduced to the concept of 3D printing at earlier ages. The prospects of 3D printing are growing, and as more people have access to this new innovation, new uses in households will emerge.[165]

The OpenReflex SLR film camera was developed for 3D printing as an open-source student project.[166]

1.7.4 Education and research

3D printing, and open source RepRap 3D printers in particular, are the latest technology making inroads into the classroom.[167][168][169] 3D printing allows students to create prototypes of items without the use of expensive tooling required in subtractive methods. Students design and produce actual models they can hold. The classroom environment allows students to learn and employ new applications for 3D printing.[170] RepRaps, for example, have already been used for an educational mobile robotics platform.[171]

Some authors have claimed that RepRap 3D printers offer an unprecedented "revolution" in STEM education.[172] The evidence for such claims comes from both the low cost ability for rapid prototyping in the classroom by students, but also the fabrication of low-cost high-quality scientific equipment from open hardware designs forming open-source labs.[173] Engineering and design principles are explored as well as architectural planning. Students recreate duplicates of museum items such as fossils and historical artifacts for study in the classroom without possibly damaging sensitive collections. Other students interested in graphic designing can construct models with complex working parts. 3D printing gives students a new perspective with topographic maps. Science students can study cross-sections of internal organs of the human body and other biological specimens. And chemistry students can explore 3D models of molecules and the relationship within chemical compounds.[174]

According to a recent paper by Kostakis et al.,[175] 3D printing and design can electrify various literacies and creative capacities of children in accordance with the spirit of the interconnected, information-based world.

Future applications for 3D printing might include creating open-source scientific equipment.[173][176]

1.7.5   Environmental use

In Bahrain, large-scale 3D printing using a sandstone-like material has been used to create unique coral-shaped structures, which encourage coral polyps to colonize and regenerate damaged reefs. These structures have a much more natural shape than other structures used to create artificial reefs, and, unlike concrete, are neither acid nor alkaline with neutral pH.[177]

1.7.6   Specialty materials

Consumer grade 3D printing has resulted in new materials that have been developed specifically for 3D printers. For example, filament materials have been developed to imitate wood, in its appearance as well as its texture. Furthermore, new technologies, such as infusing carbon fiber[178] into printable plastics, allowing for a stronger, lighter material. In addition to new structural materials that have been developed due to 3D printing, new technologies have allowed for patterns to be applied directly to 3D printed parts. Iron oxide-free Portland cement powder has been used to create architectural structures up to 9 feet in height.[179][180][181]

1.8   Legal aspects

1.8.1   Intellectual property

See also: Free hardware

3D printing has existed for decades within certain manufacturing industries where many legal regimes, including patents, industrial design rights, copyright, and trademark may apply. However, there is not much jurisprudence to say how these laws will apply if 3D printers become mainstream and individuals and hobbyist communities begin manufacturing items for personal use, for non-profit distribution, or for sale.

Any of the mentioned legal regimes may prohibit the distribution of the designs used in 3D printing, or the distribution or sale of the printed item. To be allowed to do these things, where an active intellectual property was involved, a person would have to contact the owner and ask for a licence, which may come with conditions and a price. However, many patent, design and copyright laws contain a standard limitation or exception for 'private', 'non-commercial' use of inventions, designs or works of art protected under intellectual property (IP). That standard limitation or exception may leave such private, non-commercial uses outside the scope of IP rights.

Patents cover inventions including processes, machines, manufactures, and compositions of matter and have a finite duration which varies between countries, but generally 20 years from the date of application. Therefore, if a type of wheel is patented, printing, using, or selling such a wheel could be an infringement of the patent.[182]

Copyright covers an expression[183] in a tangible, fixed medium and often lasts for the life of the author plus 70 years thereafter.[184] If someone makes a statue, they may have copyright on the look of that statue, so if someone sees that statue, they cannot then distribute designs to print an identical or similar statue.

When a feature has both artistic (copyrightable) and functional (patentable) merits, when the question has appeared in US court, the courts have often held the feature is not copyrightable unless it can be separated from the functional aspects of the item.[184] In other countries the law and the courts may apply a different approach allowing, for example, the design of a useful device to be registered (as a whole) as an industrial design on the understanding that, in case of unauthorized copying, only the non-functional features may be claimed under design law whereas any technical features could only be claimed if covered by a valid patent.

1.8.2   Gun legislation and administration

The US Department of Homeland Security and the Joint Regional Intelligence Center released a memo stating that "significant advances in three-dimensional (3D) printing capabilities, availability of free digital 3D printable files for firearms components, and difficulty regulating file sharing may present public safety risks from unqualified gun seekers who obtain or manufacture 3D printed guns," and that "proposed legislation to ban 3D printing of weapons may deter, but cannot completely prevent their production. Even if the practice is prohibited by new legislation, online distribution of these 3D printable files will be as difficult to control as any other illegally traded music, movie or software files."[185]

Internationally, where gun controls are generally tighter than in the United States, some commentators have said the impact may be more strongly felt, as alternative firearms are not as easily obtainable.[186] European officials have noted that producing a 3D printed gun would be illegal under their gun control laws,[187] and that criminals have access to other sources of weapons, but noted that as the technology improved the risks of an effect would increase.[188][189] Downloads of the plans from the UK, Germany, Spain, and Brazil were heavy.[190][191]

Attempting to restrict the distribution over the Internet of gun plans has been likened to the futility of preventing the widespread distribution of DeCSS which enabled DVD ripping.[192][193][194][195] After the US government had Defense Distributed take down the plans, they were still widely available via The Pirate Bay and other file sharing sites.[196] Some US legislators have proposed regulations on 3D printers, to prevent them being used for printing guns.[197][198] 3D printing advocates have suggested that such regulations would be futile, could cripple the 3D printing industry, and could infringe on free speech rights, with early pioneer of 3D printing Professor Hod Lipson suggesting that gunpowder could be controlled instead.[199][200][201][202][203][204][205]

1.9 Impact

Additive manufacturing, starting with today's infancy period, requires manufacturing firms to be flexible, ever-improving users of all available technologies to remain competitive. Advocates of additive manufacturing also predict that this arc of technological development will counter globalisation, as end users will do much of their own manufacturing rather than engage in trade to buy products from other people and corporations.[3] The real integration of the newer additive technologies into commercial production, however, is more a matter of complementing traditional subtractive methods rather than displacing them entirely.[206]

1.9.1 Social change

Since the 1950s, a number of writers and social commentators have speculated in some depth about the social and cultural changes that might result from the advent of commercially affordable additive manufacturing technology.[207] Amongst the more notable ideas to have emerged from these inquiries has been the suggestion that, as more and more 3D printers start to enter people's homes, so the conventional relationship between the home and the workplace might get further eroded.[208] Likewise, it has also been suggested that, as it becomes easier for businesses to transmit designs for new objects around the globe, so the need for high-speed freight services might also become less.[209] Finally, given the ease with which certain objects can now be replicated, it remains to be seen whether changes will be made to current copyright legislation so as to protect intellectual property rights with the new technology widely available.

As 3D printers became more accessible to consumers, online social platforms have developed to support the community.[210] This includes websites that allow users to access information such as how to build a 3D printer, as well as social forums that discuss how to improve 3D print quality and discuss 3D printing news, as well as social media websites that are dedicated to share 3D models.[211][212][213] RepRap is a wiki based website that was created to hold all information on 3d printing, and has developed into a community that aims to bring 3D printing to everyone. Furthermore, there are other sites such as Pinshape, Thingiverse and MyMiniFactory, which was created initially to allow users to post 3D files for anyone to print, allowing for decreased transaction cost of sharing 3D files. These websites have allowed for greater social interaction between users, creating communities dedicated around 3D printing.

Some [214][215][216] call attention to the conjunction of Commons-based peer production with 3D printing and other low-cost manufacturing techniques. The self-reinforced fantasy of a system of eternal growth can be overcome with the development of economies of scope, and here, the civil society can play an important role contributing to the raising of the whole productive structure to a higher plateau of more sustainable and customized productivity.[214] Further, it is true that many issues, problems and threats rise due to the large democratization of the means of production, and especially regarding the physical ones.[214] For instance, the recyclability of advanced nanomaterials is still questioned; weapons manufacturing could become easier; not to mention the implications on counterfeiting [217] and on IP.[218] It might be maintained that in contrast to the industrial paradigm whose competitive dynamics were about economies of scale, Commons-based peer production and 3D printing could develop economies of scope. While the advantages of scale rest on cheap global transportation, the economies of scope share infrastructure costs (intangible and tangible productive resources), taking advantage of the capabilities of the fabrication tools.[214] And following Neil Gershenfeld [219] in that "some of the least developed parts of the world need some of the most advanced technologies", Commons-based peer production and 3D printing may offer the necessary tools for thinking globally but act locally in response to certain problems and needs.

Larry Summers wrote about the "devastating consequences" of 3-D printing and other technologies (robots, artificial intelligence, etc.) for those who perform routine tasks. In his view, "already there are more American men on disability insurance than doing production work in manufacturing. And the trends are all in the wrong direction, particularly for the less skilled, as the capacity of capital embodying artificial intelligence to replace white-collar as well as blue-collar work will increase rapidly in the years ahead." Summers recommends more vigorous cooperative efforts to address the "myriad devices" (e.g. tax havens, bank secrecy, money laundering, and regulatory arbitrage) enabling the holders of great wealth to "avoid paying" income and estate taxes, and to make it more difficult to accumulate great fortunes without requiring "great social contributions" in return, including: more vigorous enforcement of anti-monopoly laws, reductions in "excessive" protection for intellectual property, greater encouragement of profit-sharing schemes that may benefit workers and give them a stake in wealth accumulation, strengthening of collective bargaining arrangements, improvements in corporate governance, strengthening of financial regulation to eliminate subsidies to financial activity, easing of land-use restrictions that may cause the real estate of the rich to keep rising in value, better training for young people and retraining for displaced workers, and increased public and private investment in infrastructure development, e.g. in energy production and transportation.[220]

Michael Spence wrote that "Now comes a ... powerful, wave of digital technology that is replacing labor in increasingly complex tasks. This process of labor substitution and disintermediation has been underway for some time in service sectors – think of ATMs, online banking, enterprise resource planning, customer relationship management, mobile payment systems, and much more. This revolution is spreading to the production of goods, where robots and 3D printing are displacing labor." In his view, the vast majority of the cost of digital technologies comes at the start, in the design of hardware (e.g. 3D printers) and, more important, in creating the software that enables machines to carry out various tasks. "Once this is achieved, the marginal cost of the hardware is relatively low (and declines as scale rises), and the marginal cost of replicating the software is essentially zero. With a huge potential global market to amortize the upfront fixed costs of design and testing, the incentives to invest [in digital technologies] are compelling." Spence believes that, unlike prior digital technologies, which drove firms to deploy underutilized pools of valuable labor around the world, the motivating force in the current wave of digital technologies "is cost reduction via the replacement of labor." For example, as the cost of 3D printing technology declines, it is "easy to imagine" that production may become "extremely" local and customized. Moreover, production may occur in response to actual demand, not anticipated or forecast demand. Spence believes that labor, no matter how inexpensive, will become a less important asset for growth and employment expansion, with labor-intensive, process-oriented manufacturing becoming less effective, and that re-localization will appear in both developed and developing countries. In his view, production will not disappear, but it will be less labor-intensive, and all countries will eventually need to rebuild their growth models around digital technologies and the human capital supporting their deployment and expansion. Spence writes that "the world we are entering is one in which the most powerful global flows will be ideas and digital capital, not goods, services, and traditional capital. Adapting to this will require shifts in mindsets, policies, investments (especially in human capital), and quite possibly models of employment and distribution."[221]

Forbes investment pundits have predicted that 3D printing may lead to a resurgence of American Manufacturing, citing the small, creative companies that comprise the current industry landscape, and the lack of the necessary complex infrastructure in typical outsource markets.[222]

1.10   See also

- List of 3D printer manufacturers

- List of notable 3D printed weapons and parts

- List of common 3D test models

- List of emerging technologies
- AstroPrint
- 3D printing marketplace
- Additive Manufacturing File Format
- Cloud manufacturing
- Computer numeric control
- Continuous Liquid Interface Production
- Digital modeling and fabrication
- 3D printer extruder
- Fusion3
- Laser cutting
- MakerBot Industries
- Magnetically assisted slip casting
- Mass customization
- Milling center
- Modular design
- Molecular assembler
- Open design
- Open source hardware
- Organ-on-a-chip
- Self-replicating machine
- Shapeways
- Thingiverse
- Volumetric printing

1.11 References

[1] Excell, Jon. "The rise of additive manufacturing". The engineer. Retrieved 2013-10-30.

[2] "3D Printer Technology – Animation of layering". Create It Real. Retrieved 2012-01-31.

[3] Jane Bird (2012-08-08). "Exploring the 3D printing opportunity". The Financial Times. Retrieved 2012-08-30.

[4] Hideo Kodama, "A Scheme for Three-Dimensional Display by Automatic Fabrication of Three-Dimensional Model," IEICE TRANSACTIONS on Electronics (Japanese Edition), vol.J64-C, No.4, pp.237-241, April 1981

[5] Hideo Kodama, "Automatic method for fabricating a three-dimensional plastic model with photo-hardening polymer," Review of Scientific Instruments, Vol. 52, No. 11, pp 1770-1773, November 1981

[6] "3D Printing: What You Need to Know". PCMag.com. Retrieved 2013-10-30.

[7] Apparatus for Production of Three-Dimensional Objects by Stereolithography (8 August 1984)

[8] Freedman, David H. "Layer By Layer." *Technology Review* 115.1 (2012): 50–53. *Academic Search Premier.* Web. 26 July 2013.

[9] Amon, C. H.; Beuth, J. L.; Weiss, L. E.; Merz, R.; Prinz, F. B. (1998). "Shape Deposition Manufacturing With Microcasting: Processing, Thermal and Mechanical Issues" (PDF). *Journal of Manufacturing Science and Engineering* **120** (3). Retrieved 2014-12-20.

[10] Beck, J.E.; Fritz, B.; Siewiorek, Daniel; Weiss, Lee (1992). "Manufacturing Mechatronics Using Thermal Spray Shape Deposition" (PDF). *Proceedings of the 1992 Solid Freeform Fabrication Symposium.* Retrieved 2014-12-20.

[11] Prinz, F. B.; Merz, R.; Weiss, Lee (1997). Ikawa, N., ed. *Building Parts You Could Not Build Before.* Proceedings of the 8th International Conference on Production Engineering. 2-6 Boundary Row, London SE1 8HN, UK: Chapman & Hall. pp. 40–44.

[12] "Google Ngram of the term additive manufacturing".

[13] GrabCAD, *GE jet engine bracket challenge*

[14] Zelinski, Peter (2014-06-02), "How do you make a howitzer less heavy?", *Modern Machine Shop*

[15] Pearce, Joshua M.; et al. (2010). "3-D Printing of Open Source Appropriate Technologies for Self-Directed Sustainable Development". *Journal of Sustainable Development* **3** (4): 17–29. doi:10.5539/jsd.v3n4p17. Retrieved 2012-01-31.

[16] Tech for Trade, 3D4D Challenge

[17] Ishengoma, Fredrick R.; Mtaho, Adam B. (2014-10-18). "3D Printing: Developing Countries Perspectives". *International Journal of Computer Applications* **104** (11): 30–34. doi:10.5120/18249-9329. ISSN 0975-8887.

[18] "Objet Connex 3D Printers". Objet Printer Solutions. Retrieved 2012-01-31.

[19] Frick, Lindsey. How to Smooth 3D-Printed Parts. Machine Design Magazine, 29 April 2014

[20] Amberlee S. Haselhuhn, Eli J. Gooding, Alexandra G. Glover, Gerald C. Anzalone, Bas Wijnen, Paul G. Sanders, Joshua M. Pearce. (2014). "Substrate Release Mechanisms for Gas Metal Arc 3-D Aluminum Metal Printing.". *3D Printing and Additive Manufacturing* **1** (4): 204–209. doi:10.1089/3dp.2014.0015.

[21] Amberlee S. Haselhuhn, Bas Wijnen, Gerald C. Anzalone, Paul G. Sanders, Joshua M. Pearce, In Situ Formation of Substrate Release Mechanisms for Gas Metal Arc Weld Metal 3-D Printing. *Journal of Materials Processing Technology.* 226, pp. 50–59 (2015) DOI: 10.1016/j.jmatprotec.2015.06.038 10.1016/j.jmatprotec.2015.06.038

[22] FDM is a proprietary term owned by Stratasys. All 3-D printers that are not Stratasys machines and use a fused filament process are referred to as or fused filament fabrication (FFF).

[23] Sherman, Lilli Manolis (November 15, 2007). "A whole new dimension – Rich homes can afford 3D printers". *The Economist.*

[24] Wohlers, Terry. "Factors to Consider When Choosing a 3D Printer (WohlersAssociates.com, Nov/Dec 2005)".

[25] www.3ders.org (2012-09-25). "Casting aluminum parts directly from 3D printed PLA parts". 3ders.org. Retrieved 2013-10-30.

[26] "Affordable 3D Printing with new Selective Heat Sintering (SHS™) technology". blueprinter.

[27] Chee Kai Chua; Kah Fai Leong; Chu Sing Lim (2003). *Rapid Prototyping.* World Scientific. p. 124. ISBN 978-981-238-117-0.

[28] Frick, Lindsey. Aluminum-powder DMLS-printed part finishes race first. Machine Design Magazine, 3 March 2014

[29] Deckard, C., "Method and apparatus for producing parts by selective sintering", U.S. Patent 4,863,538, filed October 17, 1986, published September 5, 1989.

[30] Housholder, R., "Molding Process", U.S. Patent 4,247,508, filed December 3, 1979, published January 27, 1981.

[31] Hiemenz, Joe. "Rapid prototypes move to metal components (EE Times, 3/9/2007)".

[32] "Rapid Manufacturing by Electron Beam Melting". SMU.edu.

[33] "3D Printer Uses Standard Paper", "Rapid Today", May, 2008

[34] U.S. Patent 4,575,330

[35] NSF JTEC/WTEC Panel Report-RPA

[36] Beaumont Newhall (May 1958) "Photosculpture," *Image*, **7** (5) : 100–105

[37] François Willème, "Photo-sculpture," U.S. Patent no. 43,822 (August 9, 1864). Available on-line at: U.S. Patent 43,822

[38] François Willème (May 15, 1861) "La sculpture photographique", *Le Moniteur de la photographie*, p. 34.

[39] "EnvisionTEC Perfactory". EnvisionTEC.

[40] Johnson, R. Colin. "Cheaper avenue to 65 nm? (EE Times, 3/30/2007)".

[41] "The World's Smallest 3D Printer". TU Wien. 12 September 2011.

[42] "3D-printing multi-material objects in minutes instead of hours". Kurzweil Accelerating Intelligence. November 22, 2013.

[43] "3D Printing: Challenges and Opportunities for International Relations". *Transcript.* Council on Foreign Relations. October 23, 2013. Retrieved 2013-10-30.

[44] Kalish, Jon. "A Space For DIY People To Do Their Business (NPR.org, November 28, 2010)". Retrieved 2012-01-31.

[45] Jones, R., Haufe, P., Sells, E., Iravani, P., Olliver, V., Palmer, C., & Bowyer, A. (2011). Reprap-- the replicating rapid prototyper. Robotica, 29(1), 177-191.

[46] "Open source 3D printer copies itself". Computerworld New Zealand. 2008-04-07. Retrieved 2013-10-30.

[47] RepRap blog 2009 visited 2/26/2014

[48] An Inexpensive Way to Print Out Metal Parts - The New York Times

[49] Gerald C. Anzalone, Chenlong Zhang, Bas Wijnen, Paul G. Sanders and Joshua M. Pearce, " Low-Cost Open-Source 3-D Metal Printing" *IEEE Access*, 1, pp.803-810, (2013). doi: 10.1109/ACCESS.2013.2293018

[50] Disruptions: 3-D Printing Is on the Fast Track – NYTimes.com

[51] www.3ders.org. "3D printers list with prices". 3ders.org. Retrieved 2013-10-30.

[52] New Scientist magazine: Desktop fabricator may kick-start home revolution, 9 January 2007

[53] "3D printer by Saskatchewan man gets record crowdsourced cash". Saskatchewan: CBC News. 6 November 2013. Retrieved 8 November 2013.

[54] "Rapide One – Affordable Professional Desktop 3D Printer by Rapide 3D". Indiegogo. December 2, 2013. Retrieved 20 January 2014.

[55] Pogue, David. "A Review Of The 3Doodler Pen, Which Raised Over $2 Million On Kickstarter". Yahoo Tech. Retrieved 13 March 2014

[56] Dorrier, Jason. "Kickstarter 3Doodler 3D Printing Pen Nothing of the Sort – But Somehow Raises $2 Million". Singularity Hub. Retrieved 13 March 2014

[57] Wittbrodt, B. T.; Glover, A. G.; Laureto, J.; Anzalone, G. C.; Oppliger, D.; Irwin, J. L.; Pearce, J. M. (2013). "Life-cycle economic analysis of distributed manufacturing with open-source 3-D printers". *Mechatronics* **23** (6): 713. doi:10.1016/j.mechatronics.2013.06.0

[58] Kreiger, M.; Pearce, J. M. (2013). "Environmental Life Cycle Analysis of Distributed Three-Dimensional Printing and Conventional Manufacturing of Polymer Products". *ACS Sustainable Chemistry & Engineering*: 131002082320002. doi:10.1021/sc400093k.

[59] Christian Baechler, Matthew DeVuono, and Joshua M. Pearce, "Distributed Recycling of Waste Polymer into RepRap Feedstock". *Rapid Prototyping Journal,* **19** (2), pp. 118-125 (2013). DOI:10.1108/13552541311302978

[60] Kreiger, M., Anzalone, G. C., Mulder, M. L., Glover, A., & Pearce, J. M. (2013). Distributed Recycling of Post-Consumer Plastic Waste in Rural Areas. MRS Online Proceedings Library, 1492, mrsf12-1492

[61] See for example the Rostock

[62] Vandendriessche, Pieter-Jan. "delta 3D printer accuracy".

[63] Titsch, Mike (July 11, 2013). "MatterHackers Opens 3D Printing Store and Releases MatterControl 0.7.6". Retrieved November 30, 2013.

[64] "Hoosier Daddy – The Largest Delta 3D Printer In the World". *3D Printer World* (Punchbowl Media). 23 July 2014. Retrieved 28 September 2014.

[65] McKenna, Beth (26 April 2014). "The Next Big Thing in 3-D Printing: Big Area Additive Manufacturing, or BAAM". *The Motley Fool*. Retrieved 28 September 2014.

[66] J. L. Wasserman; et al. (2008). "Fabrication of One-Dimensional Programmable-Height Nanostructures via Dynamic Stencil Deposition". *Review of Scientific Instruments* **79**: 073909. arXiv:0802.1848. doi:10.1063/1.2960573.

[67] patel, Prachi (5 March 2013). "Micro 3-D Printer Creates Tiny Structures in Seconds". *MIT Technology Review*.

[68] "Print me a Stradivarius – How a new manufacturing technology will change the world". Economist Technology. 2011-02-10. Retrieved 2012-01-31.

[69] Zelinski, Peter (2014-06-25), "Video: World's largest additive metal manufacturing plant", *Modern Machine Shop*

[70] Sherman, Lilli Manolis. "3D Printers Lead Growth of Rapid Prototyping (Plastics Technology, August 2004)". Retrieved 2012-01-31.

[71] "3D printing: 3D printing scales up". The Economist. 2013-09-07. Retrieved 2013-10-30.

[72] Development of a Three-Dimensional Printed, Liquid-Cooled Nozzle for a Hybrid Rocket Motor, Nick Quigley and James Evans Lyne, Journal of Propulsion and Power, Vol. 30, No. 6 (2014), pp. 1726-1727.

[73] Vincent & Earls 2011

[74] G. Anzalone, B. Wijnen, Joshua M. Pearce , (2015) "Multi-material additive and subtractive prosumer digital fabrication with a free and open-source convertible delta RepRap 3-D printer", *Rapid Prototyping Journal*, 21(5), pp.506 - 519. doi:10.1108/RPJ-09-2014-0113

[75] Felix Bopp (2010). *Future Business Models by Additive Manufacturing*. Verlag. ISBN 3836685086. Retrieved 4 July 2014.

[76] "3D Hubs: Like Airbnb For 3D Printers". gizmodo. Retrieved 2014-07-05.

[77] Sterling, Bruce (2011-06-27). "Spime Watch: Dassault Systèmes' 3DVIA and Sculpteo (Reuters, June 27, 2011)". *Wired*. Archived from the original on 15 April 2014. Retrieved 2012-01-31.

[78] Vance, Ashlee (2011-01-12). "The Wow Factor of 3-D Printing (The New York Times, January 12, 2011)". Retrieved 2012-01-31.

[79] "The action doll you designed, made real". makie.me. Retrieved January 18, 2013.

[80] "Cubify — Express Yourself in 3D". myrobotnation.com. Retrieved 2014-01-25.

[81] "Turn Your Baby's Cry Into an iPhone Case". Bloomberg Businessweek. 2012-03-10. Retrieved 2013-02-20

[82] "Nokia backs 3D printing for mobile phone cases". BBC News Online. 2013-02-18. Retrieved 2013-02-20

[83] Wohlers Report 2009, State of the Industry Annual Worldwide Progress Report on Additive Manufacturing, Wohlers Associates, ISBN 978-0-9754429-5-1

[84] Hopkinson, N & Dickens, P 2006, 'Emerging Rapid Manufacturing Processes', in Rapid Manufacturing; An industrial revolution for the digital age, Wiley & Sons Ltd, Chichester, W. Sussex

[85] Symes, M. D.; Kitson, P. J.; Yan, J.; Richmond, C. J.; Cooper, G. J. T.; Bowman, R. W.; Vilbrandt, T.; Cronin, L. (2012). "Integrated 3D-printed reactionware for chemical synthesis and analysis". *Nature Chemistry* **4** (5): 349–354. doi:10.1038/nchem.1313. PMID 22522253.

[86] "Hydrocolloid Printing", Cornell Creative, 2012

[87] A Guide to All the Food That's Fit to 3D Print (So Far)

[88] "Foodini 3D Printer Cooks Up Meals Like the Star Trek Food Replicator". Retrieved 27 January 2015.

[89] "NASA - 3D Printing: Food in Space". *www.nasa.gov*. Retrieved 2015-09-30.

[90] ted.com, Lee Cronin: Print your own medicine

[91] "RFA-HD-15-023: Use of 3-D Printers for the Production of Medical Devices (R43/R44)". *grants.nih.gov*. Retrieved 2015-09-30.

[92] "3D Printed Clothing Becoming a Reality". Resins Online. 2013-06-17. Retrieved 2013-10-30.

[93] Michael Fitzgerald (2013-05-28). "With 3-D Printing, the Shoe Really Fits". MIT Sloan Management Review. Retrieved 2013-10-30.

[94] "3D Custom Eyewear The Next Focal Point For 3D Printing". Rakesh Sharma. 2013-09-10. Retrieved 2013-09-10.

[95] "Koenigsegg One:1 Comes With 3D Printed Parts". Business Insider. Retrieved 2014-05-14.

[96] tecmundo.com.br/ *Conheça o Urbee, primeiro carro a ser fabricado com uma impressora 3D*

[97] The "Urbee" 3D-Printed Car: Coast to Coast on 10 Gallons Truthout

[98] 3D Printed Car Creator Discusses Future of the Urbee

[99] Zitun, Yoav (July 27, 2015). "The 3D printer revolution comes to the IAF". Ynet News. Retrieved 2015-09-29.

[100] Knippers, E. "Architecture | Leapfrog 3D Printers". *www.lpfrg.com*. Retrieved 2015-09-29.

[101] "Concept Modeling, Realistic 3D Printed Models | Stratasys". *www.stratasys.com*. Retrieved 2015-09-29.

[102] Khoshnevis, B.; Russell, R.; Kwon, Hongkyu; Bukkapatnam, S. (2001-09-01). "Crafting large prototypes". *IEEE Robotics Automation Magazine* **8** (3): 33–42. doi:10.1109/100.956812. ISSN 1070-9932.

[103] Greenberg, Andy (2012-08-23). "'Wiki Weapon Project' Aims To Create A Gun Anyone Can 3D-Print At Home". Forbes. Retrieved 2012-08-27.

[104] Poeter, Damon (2012-08-24). "Could a 'Printable Gun' Change the World?". PC Magazine. Retrieved 2012-08-27.

[105] "Blueprints for 3-D printer gun pulled off website". statesman.com. May 2013. Retrieved 2013-10-30.

[106] Samsel, Aaron. "3D Printers, Meet Othermill: A CNC machine for your home office (VIDEO)". Guns.com. Retrieved 2013-10-30.

[107] "The Third Wave, CNC, Stereolithography, and the end of gun control". Popehat. Retrieved 2013-10-30.

[108] Rosenwald, Michael S. (2013-02-25). "Weapons made with 3-D printers could test gun-control efforts". *Washington Post*.

[109] "Making guns at home: Ready, print, fire". The Economist. 2013-02-16. Retrieved 2013-10-30.

[110] Rayner, Alex (6 May 2013). "3D-printable guns are just the start, says Cody Wilson". *The Guardian* (London).

[111] Manjoo, Farhad (2013-05-08). "3-D-printed gun: Yes, it will be possible to make weapons with 3-D printers. No, that doesn't make gun control futile". Slate.com. Retrieved 2013-10-30.

[112] Franzen, Carl. "3D-printed gun maker in Japan sentenced to two years in prison".

[113] "Transplant jaw made by 3D printer claimed as first". BBC. 2012-02-06.

[114] Rob Stein (2013-03-17). "Doctors Use 3-D Printing To Help A Baby Breathe". NPR.

[115] Moore, Calen (11 February 2014). "Surgeons have implanted a 3-D-printed pelvis into a U.K. cancer patient". fiercemedicaldevices.com. Retrieved 4 March 2014.

[116] Keith Perry (12 March 2014). "Man makes surgical history after having his shattered face rebuilt using 3D printed parts". London: The Daily Telegraph. Retrieved 12 March 2014.

[117] Research into Julie Williams, 3D-Bioprinting may soon produce transplantable human tissues Mar.6, 2014

[118] BBC News (October 2014). "Inverness girl Hayley Fraser gets 3D-printed hand", BBC News, 01 October 2014. Retrieved 02 October 2014.

[119] "3D-Printed Foot Lets Crippled Duck Walk Again".

[120] Pleasance, Chris (18 August 2014). "Puppy power: Chihuahua born without front legs is given turbo-charged makeover after being fitted with 3D printed body harness and a set of skateboard wheels". The Daily Mail. Retrieved 2014-08-21.

[121] Flaherty, Joseph (2013-07-30). "So Cute: Hermit Crabs Strut in Stylish 3-D Printed Shells". *Wired*.

[122] Weintraub, Arlene (20 March 2015). "3D Systems preps for global launch of 'printed' knee implants for dogs". fierceanimalhealth.com. Retrieved 13 April 2015.

[123] Saxena, Varun (2 February 2015). "FDA clears 3-D printed device for minimally invasive foot surgery". fiercemedicaldevices.com. Retrieved 14 April 2015.

[124] "3D-printed sugar network to help grow artificial liver", BBC, 2 July 2012

[125] "Invetech helps bring bio-printers to life". *Australian Life Scientist*. Westwick-Farrow Media. December 11, 2009. Retrieved December 31, 2013.

[126] "Building body parts with 3D printing", The Engineer, 24 May 2010

[127] Silverstein, Jonathan. "'Organ Printing' Could Drastically Change Medicine (ABC News, 2006)". Retrieved 2012-01-31.

[128] Dan Thomas, Engineering Ourselves – The Future Potential Power of 3D-Bioprinting?, engineering.com, March 25, 2014

[129] The Diplomat (2013-08-15). "Chinese Scientists Are 3D Printing Ears and Livers – With Living Tissue". *Tech Biz*. The Diplomat. Retrieved 2013-10-30.

[130] "How do they 3D print kidney in China". Retrieved 2013-10-30.

[131] "Mish's Global Economic Trend Analysis: 3D-Printing Spare Human Parts; Ears and Jaws Already, Livers Coming Up ; Need an Organ? Just Print It". Globaleconomicanalysis.blogspot.co.uk. 2013-08-18. Retrieved 2013-10-30.

[132] Palmer, Eric (3 August 2015). "Company builds plant for 3DP pill making as it nails first FDA approval". fiercepharmamanufacturing.com. Retrieved 4 August 2015.

[133] The Almost Completely Open Source Laptop Goes on Sale

[134] Robots And 3D Printing

[135] Why to Use 3D Printers and the Best 3D Printers To Build Your Own Robot

[136] Printoo: Giving Life to Everyday Objects (paper-thin, flexible Arduino-compatible modules)

[137] 3&DBot: An Arduino 3D printer-robot with wheels

[138] A lesson in building a custom 3D printed humanoid robot

[139] Hays, Brooks (2014-12-19). "NASA just emailed the space station a new socket wrench". Retrieved 2014-12-20.

[140] Brabaw, Kasandra (2015-01-30). "Europe's 1st Zero-Gravity 3D Printer Headed for Space". Retrieved 2015-02-01.

[141] Wood, Anthony (2014-11-17). "POP3D to be Europe's first 3D printer in space". Retrieved 2015-02-01.

[142] Raval, Siddharth (2013-03-29). "SinterHab: A Moon Base Concept from Sintered 3D-Printed Lunar Dust". *Space Safety Magazine*. Retrieved 2013-10-15.

[143] "The World's First 3D-Printed Building Will Arrive In 2014". *TechCrunch*. 2012-01-20. Retrieved 2013-02-08.

[144] Diaz, Jesus (2013-01-31). "This Is What the First Lunar Base Could Really Look Like". *Gizmodo*. Retrieved 2013-02-01.

[145] "The RepRap's Heritage".

[146] Kelly, Heather (July 31, 2013). "Study: At-home 3D printing could save consumers "thousands"". CNN.

[147] Séquin, C. H. (2005). "Rapid prototyping". *Communications of the ACM* **48** (6): 66. doi:10.1145/1064830.1064860.

[148] Guth, Robert A. "How 3-D Printing Figures To Turn Web Worlds Real (The Wall Street Journal, December 12, 2007)" (PDF). Retrieved 2012-01-31.

[149] iPad iPhone Android TIME TV Populist The Page (2008-04-03). ""Bathsheba Grossman's Quin.MGX for Materialise" listed in Time Magazine's Design 100". Time.com. Retrieved 2013-10-30.

[150] Williams, Holly (2011-08-28). "Object lesson: How the world of decorative art is being revolutionised by 3D printing (The Independent, 28 August 2011)". London. Retrieved 2012-01-31.

[151] Bennett, Neil (November 13, 2013). "How 3D printing is helping doctors mend you better". TechAdvisor.

[152] "Custom Bobbleheads". Retrieved 13 January 2015.

[153] "3D-print your face in chocolate for that special Valentine's Day gift". The Guardian. 25 January 2013.

[154] Cignoni, P.; Scopigno, R. (2008). "Sampled 3D models for CH applications". *Journal on Computing and Cultural Heritage* **1**: 1. doi:10.1145/1367080.1367082.

[155] DiSalvo, C (2009). "Design and the Construction of Publics". *Design Issues*. 1 **25**: 48. doi:10.1162/desi.2009.25.1.48.

[156] Ratto, M. & Ree, R. (2012). "Materializing information: 3D printing and social change.". *First Monday* **17** (7).

[157] Ratto, Matt (2011). "Open Design and Critical Making". *Open Design Now: Why Design Cannot Remain Exclusive.*

[158] Lukens, Jonathan. "Speculative Design and Technological Fluency". *International Journal of Learning and Media* **3**: 23–39. doi:10.1162/ijlm_a_00080.

[159] Pandey, S.; Gupta, B.; Nahata, A. (2013). "Complex Geometry Plasmonic Terahertz Waveguides Created via 3D Printing". *Cleo: 2013*. pp. CTh1K.CTh12. doi:10.1364/CLEO_SI.2013.CTh1K.2. ISBN 978-1-55752-972-5.

[160] ewilhelm. "3D printed clock and gears". Instructables.com. Retrieved 2013-10-30.

[161] 23/01/2012 (2012-01-23). "Successful Sumpod 3D printing of a herringbone gear". 3d-printer-kit.com. Retrieved 2013-10-30.

[162] Search engine for 3D printable models, "backscratcher", etc.

[163] New Scientist magazine: Make your own drugs with a 3D printer, 17 April 2012

[164] Cronin, Lee (2012-04-17). "3D printer developed for drugs" (video interview [5:21]). Glasgow University: BBC News Online. Retrieved 2013-03-06.

[165] D'Aveni, Richard. "3-D Printing Will Change the World". *Harvard Business Review*. Retrieved October 8, 2014.

[166] "3D printable SLR brings whole new meaning to "digital camera"". Gizmag.com. Retrieved 2013-10-30.

[167] Schelly, C., Anzalone, G., Wijnen, B., & Pearce, J. M. (2015). Open-source 3-D printing Technologies for education: Bringing Additive Manufacturing to the Classroom. *Journal of Visual Languages & Computing.*

[168] Grujović, N., Radović, M., Kanjevac, V., Borota, J., Grujović, G., & Divac, D. (2011, September). 3D printing technology in education environment. *In 34th International Conference on Production Engineering* (pp. 29-30).

[169] Mercuri, R., & Meredith, K. (2014, March). An educational venture into 3D Printing. In Integrated STEM Education Conference (ISEC), 2014 IEEE (pp. 1-6). IEEE.

[170] Students Use 3D Printing to Reconstruct Dinosaurs - YouTube

[171] Gonzalez-Gomez, J., Valero-Gomez, A., Prieto-Moreno, A., & Abderrahim, M. (2012). A new open source 3d-printable mobile robotic platform for education. In *Advances in autonomous mini robots* (pp. 49-62). Springer Berlin Heidelberg.

[172] J. Irwin, J.M. Pearce, D. Opplinger, and G. Anzalone. The RepRap 3-D Printer Revolution in STEM Education, *121st ASEE Annual Conference and Exposition, Indianapolis, IN.* Paper ID #8696 (2014).

[173] Zhang, C.; Anzalone, N. C.; Faria, R. P.; Pearce, J. M. (2013). De Brevern, Alexandre G, ed. "Open-Source 3D-Printable Optics Equipment". *PLoS ONE* **8** (3): e59840. doi:10.1371/journal.pone.0059840. PMC 3609802. PMID 23544104.

[174] 3D Printing in the Classroom to Accelerate Adoption of Technology. On 3D Printing

[175] Kostakis, V.; Niaros, V.; Giotitsas, C. (2014): *Open source 3D printing as a means of learning: An educational experiment in two high schools in Greece.* In: Telematics and Informatics

[176] Pearce, Joshua M. 2012. "Building Research Equipment with Free, Open-Source Hardware." *Science* **337** (6100): 1303–1304

[177] "Underwater City: 3D Printed Reef Restores Bahrain's Marine Life". ptc.com. 2013-08-01. Retrieved 2013-10-30.

[178] Eitel, Elisabeth. MarkForged: $5,000 3D printer prints carbon-fiber parts. Machine Design Magazine, 7 March 2014

[179] "Researchers at UC Berkeley Create Bloom First Ever 3D-printed Cement Structure That Stands 9 Feet Tall". cbs sanfrancisco. 6 March 2015. Retrieved 23 April 2015.

[180] Chino, Mike (9 March 2015). "UC Berkeley unveils 3D-printed "Bloom" building made of powdered cement". Retrieved 23 April 2015.

[181] Fixsen, Anna (6 March 2015). "Print it Real Good: First Powder-Based 3D Printed Cement Structure Unveiled". Retrieved 23 April 2015.

[182] 3D Printing Technology Insight Report, 2014, patent activity involving 3D-Printing from 1990-2013, accessed 2014-06-10

[183] Clive Thompson on 3-D Printing's Legal Morass. Wired, Clive Thompson 05.30.12 1:43 PM

[184] Weinberg, Michael (January 2013). "What's the Deal with copyright and 3D printing?" (PDF). Institute for Emerging Innovation. Retrieved 2013-10-30.

[185] "Homeland Security bulletin warns 3D-printed guns may be 'impossible' to stop". Fox News. 2013-05-23. Retrieved 2013-10-30.

[186] Cochrane, Peter (2013-05-21). "Peter Cochrane's Blog: Beyond 3D Printed Guns". TechRepublic. Retrieved 2013-10-30.

[187] Gilani, Nadia (2013-05-06). "Gun factory fears as 3D blueprints put online by Defense Distributed | Metro News". Metro.co.uk. Retrieved 2013-10-30.

[188] "Liberator: First 3D-printed gun sparks gun control controversy". Digitaljournal.com. Retrieved 2013-10-30.

[189] "First 3D Printed Gun 'The Liberator' Successfully Fired". IBTimes UK. 2013-05-07. Retrieved 2013-10-30.

[190] "US demands removal of 3D printed gun blueprints". neurope.eu. Retrieved 2013-10-30.

[191] "España y EE.UU. lideran las descargas de los planos de la pistola de impresión casera". ElPais.com. 2013-05-09. Retrieved 2013-10-30.

[192] "Controlled by Guns". Quiet Babylon. 2013-05-07. Retrieved 2013-10-30.

[193] "3dprinting". Joncamfield.com. Retrieved 2013-10-30.

[194] "State Dept Censors 3D Gun Plans, Citing 'National Security'". News.antiwar.com. 2013-05-10. Retrieved 2013-10-30.

[195] "Wishful Thinking Is Control Freaks' Last Defense Against 3D-Printed Guns". Reason.com. 2013-05-08. Retrieved 2013-10-30.

[196] Lennard, Natasha (2013-05-10). "The Pirate Bay steps in to distribute 3-D gun designs". Salon.com. Archived from the original on 2013-05-19. Retrieved 2013-10-30.

[197] "Sen. Leland Yee Proposes Regulating Guns From 3-D Printers". CBS Sacramento. 2013-05-08. Retrieved 2013-10-30.

[198] Schumer Announces Support For Measure To Make 3D Printed Guns Illegal

[199] "Four Horsemen of the 3D Printing Apocalypse". Makezine.com. 2011-06-30. Retrieved 2013-10-30.

[200] Ball, James (10 May 2013). "US government attempts to stifle 3D-printer gun designs will ultimately fail". *The Guardian* (London).

[201] Gadgets (2013-01-18). "Like It Or Not, 3D Printing Will Probably Be Legislated". TechCrunch. Retrieved 2013-10-30.

[202] Klimas, Liz (2013-02-19). "Engineer: Don't Regulate 3D Printed Guns, Regulate Explosive Gun Powder Instead". TheBlaze.com. Retrieved 2013-10-30.

[203] Beckhusen, Robert (2013-02-15). "3-D Printing Pioneer Wants Government to Restrict Gunpowder, Not Printable Guns | Danger Room". Wired.com. Retrieved 2013-10-30.

[204] Bump, Philip (2013-05-10). "How Defense Distributed Already Upended the World". The Atlantic Wire. Archived from the original on 2013-05-19. Retrieved 2013-10-30.

[205] "News". European Plastics News. Retrieved 2013-10-30.

[206] Albert 2011

[207] "Confronting a New 'Era of Duplication'? 3D Printing, Replicating Technology and the Search for Authenticity in George O. Smith's Venus Equilateral Series". Durham University. Retrieved July 21, 2013.

[208] "Materializing information: 3D printing and social change". Retrieved January 13, 2014.

[209] "Additive Manufacturing: A supply chain wide response to economic uncertainty and environmental sustainability" (PDF). Retrieved January 11, 2014.

[210] "Materializing information: 3D printing and social change". Retrieved March 30, 2014.

[211] "RepRap Options". Retrieved March 30, 2014.

[212] "3D Printing". Retrieved March 30, 2014.

[213] "Thingiverse". Retrieved March 30, 2014.

[214] Kostakis, V. (2013): *At the Turning Point of the Current Techno-Economic Paradigm: Commons-Based Peer Production, Desktop Manufacturing and the Role of Civil Society in the Perezian Framework.* . In: TripleC, 11(1), 173 - 190.

[215] Kostakis, V.; Papachristou, M. (2014): *Commons-based peer production and digital fabrication: The case of a RepRap-based, Lego-built 3D printing-milling machine.* In: Telematics and Informatics, 31(3), 434 - 443

[216] Kostakis, V; Fountouklis, M; Drechsler, W. (2013): *Peer Production and Desktop Manufacturing: The Case of the Helix-T Wind Turbine Project.* . In: Science, Technology & Human Values, 38(6), 773 - 800.

[217] Campbell, Thomas, Christopher Williams, Olga Ivanova, and Banning Garrett. (2011): *Could 3D Printing Change the World? Technologies, Potential, and Implications of Additive Manufacturing.* Washington: Atlantic Council of the United States

[218] Bradshaw, Simon, Adrian Bowyer, and Patrick Haufe (2010): *The Intellectual Property Implications of Low-Cost 3D Printing.* In: SCRIPTed 7

[219] Gershenfeld, Neil (2007): *FAB: The Coming Revolution on your Desktop: From Personal Computers to Personal Fabrication.* Cambridge: Basic Books, p. 13-14

[220] Larry Summers, The Inequality Puzzle, *Democracy: A Journal of Ideas*, Issue #32, Spring 2014

[221] Michael Spence, Labor's Digital Displacement (2014-05-22), *Project Syndicate*

[222] Can 3D Printing Reshape Manufacturing In America?, Forbes.com 17 June 2014, retrieved 11 Aug 2014

1.12   Further reading

- "3D Printing Features and Usage Shootout". RS Components.

- "Results of Make Magazine's 2015 3D Printer Shootout". docs.google.com. Retrieved 1 June 2015.

- "Evaluation Protocol for Make Magazine's 2015 3D Printer Shootout". makezine.com. Retrieved 1 June 2015.

- Vincent; Earls, Alan R. (February 2011). "Origins: A 3D Vision Spawns Stratasys, Inc.". *Today's Machining World* (Oak Forest, Illinois, USA: Screw Machine World Inc.) **7** (1): 24–25.

- "Heat Beds in 3D Printing – Advantages and Equipment". *Boots Industries*. Retrieved 7 September 2015.

- Albert, Mark [Editor in Chief] (17 January 2011). "Subtractive plus additive equals more than (− + + = >): subtractive and additive processes can be combined to develop innovative manufacturing methods that are superior to conventional methods ['Mark: My Word' column – Editor's Commentary]". *Modern Machine Shop* (Cincinnati, Ohio, USA: Gardner Publications Inc.) **83** (9): 14.

- Stephens, B.; Azimi, P.; El Orch, Z.; Ramos, T. (2013). "Ultrafine particle emissions from desktop 3D printers". *Atmospheric Environment* **79**: 334. doi:10.1016/j.atmosenv.2013.06.050.

- Easton, Thomas A. (November 2008). "The 3D Trainwreck: How 3D Printing Will Shake Up Manufacturing". *Analog* **128** (11): 50–63.

- Wright, Paul K. (2001). *21st Century Manufacturing*. New Jersey: Prentice-Hall Inc.

1.13 External links

- Rapid prototyping websites at DMOZ

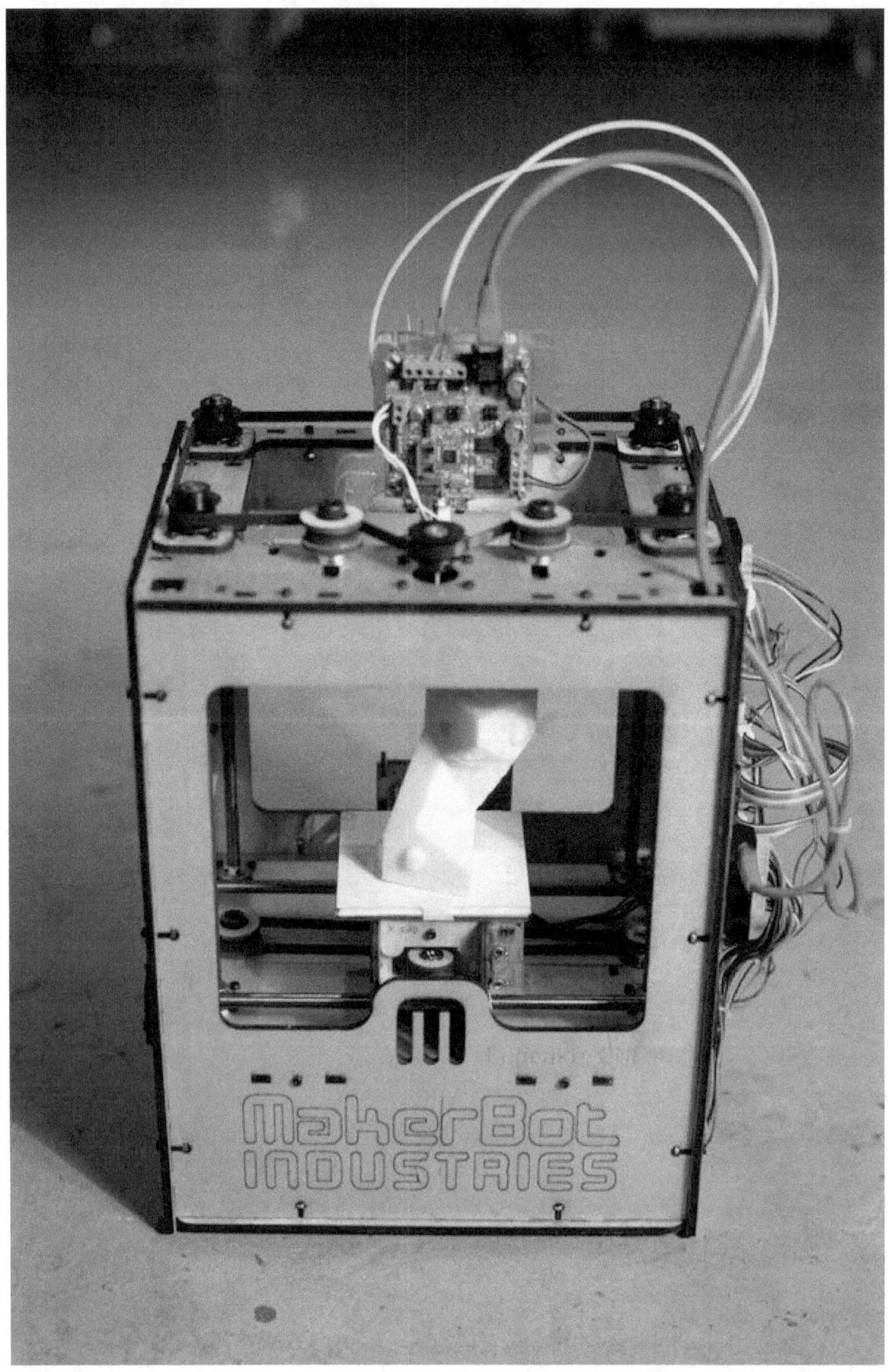

A MakerBot 3d Printer

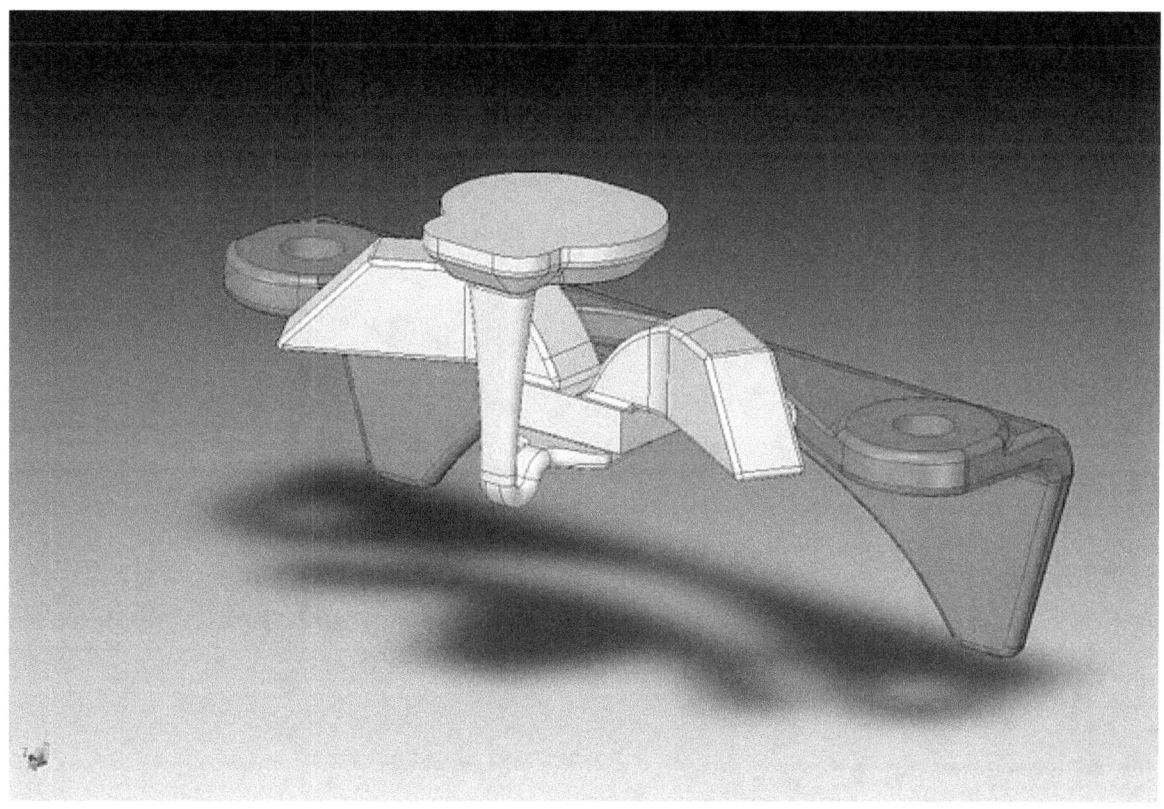

CAD model used for 3D printing

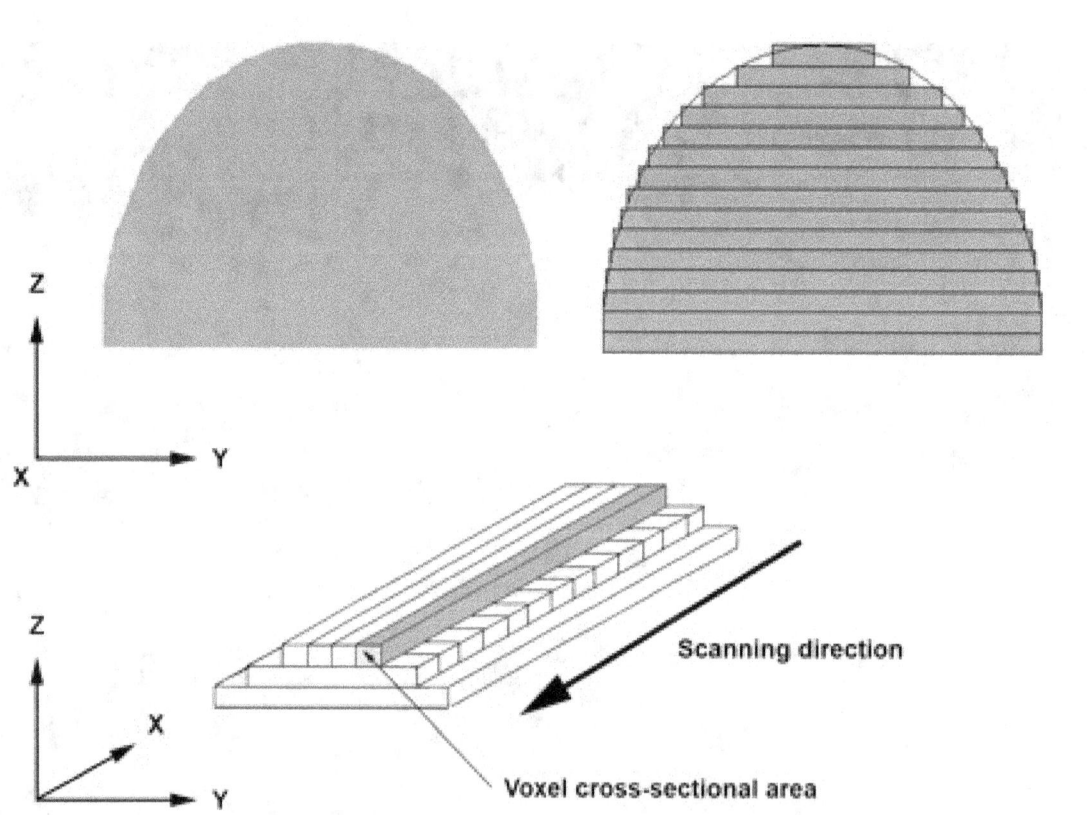

3D model slicing

Timelapse video of a hyperboloid object (designed by George W. Hart) made of PLA using a RepRap "Prusa Mendel" 3 printer for molten polymer deposition

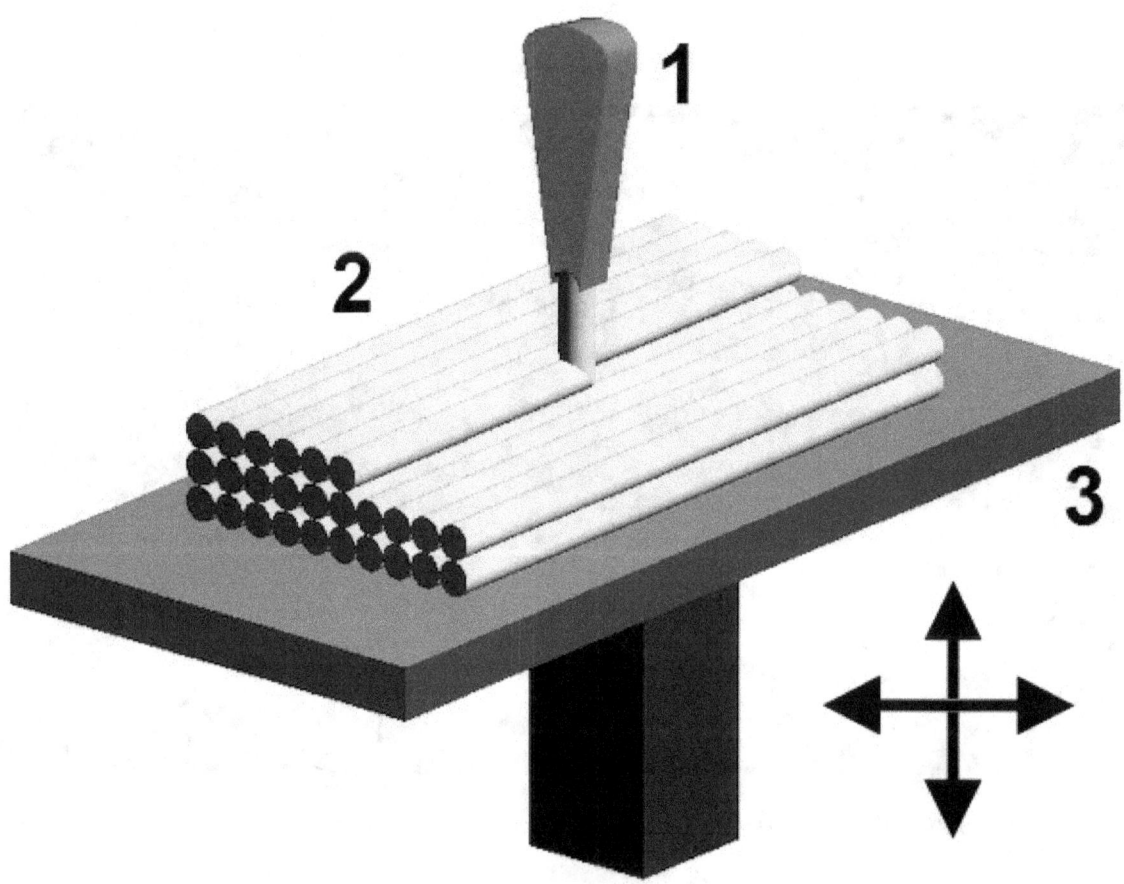

Fused deposition modeling: 1 – nozzle ejecting molten material, 2 – deposited material (modeled part), 3 – controlled movable table

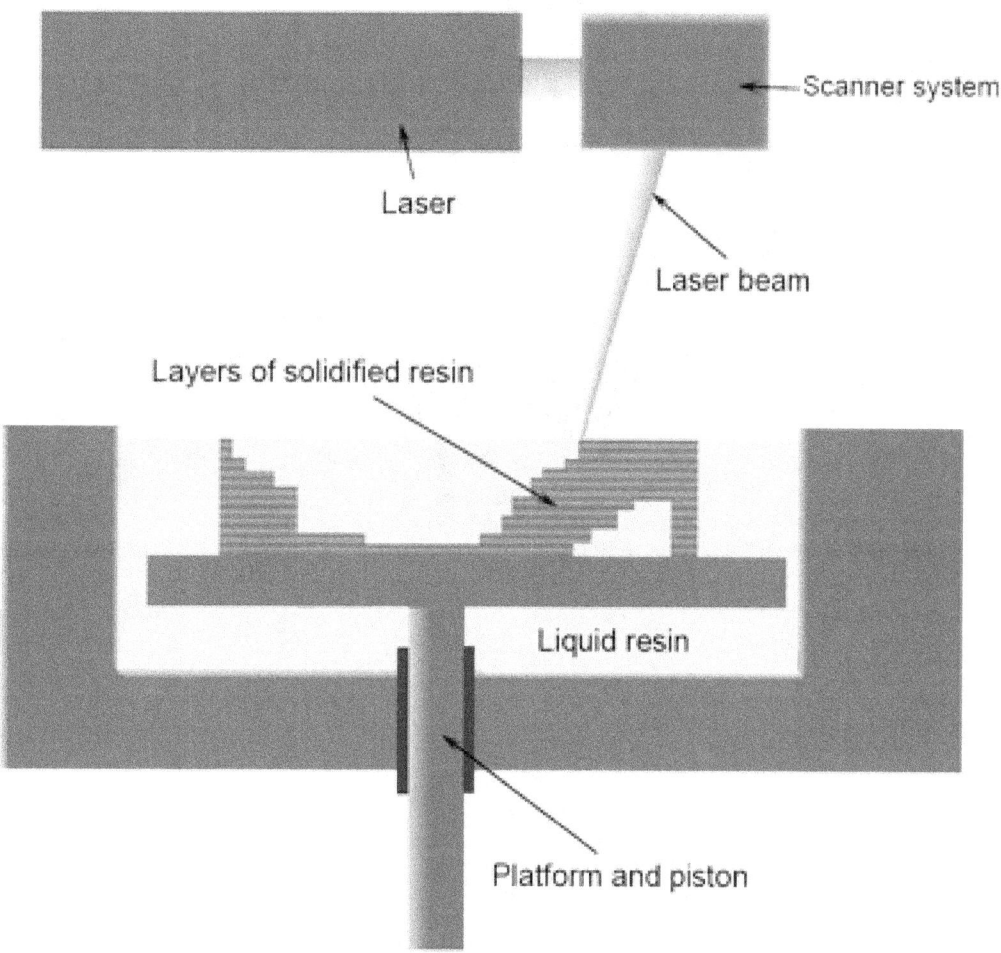

Stereolithography apparatus

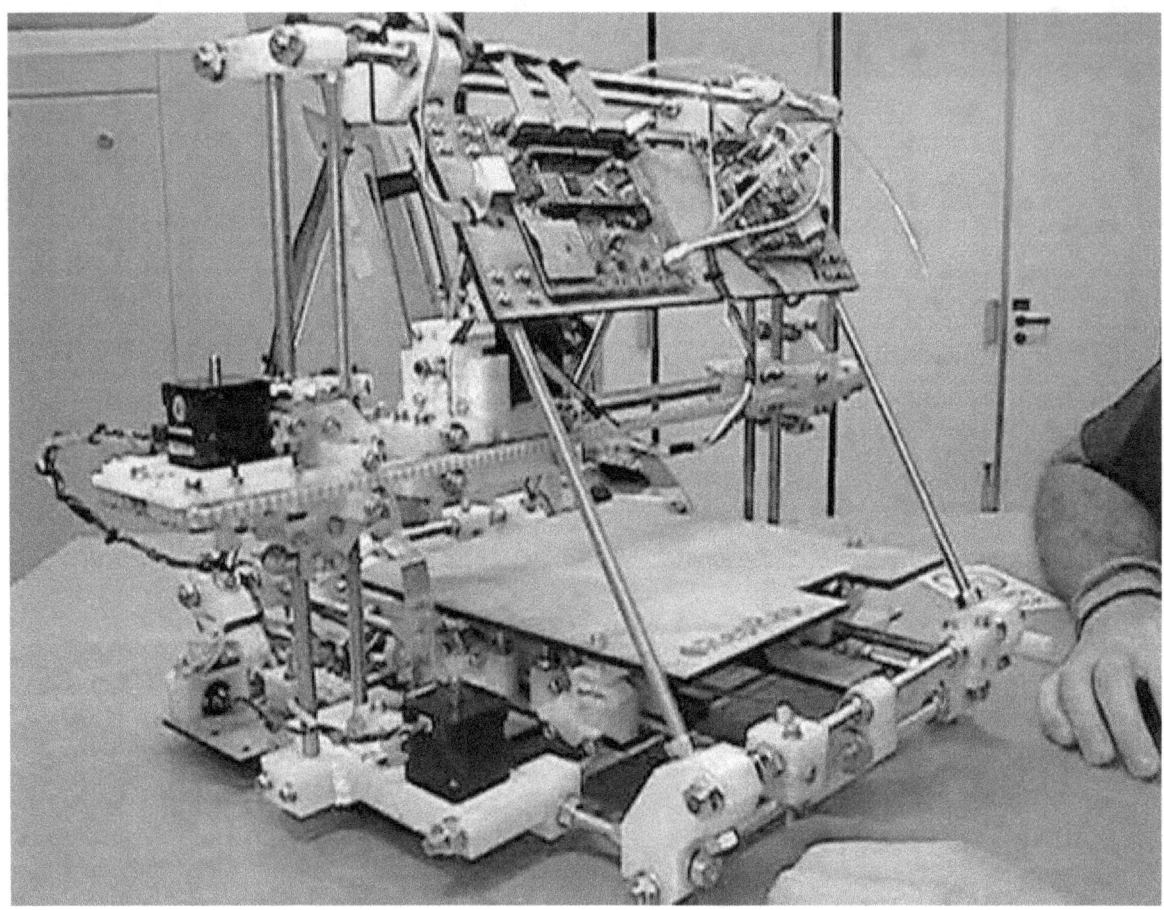

RepRap version 2.0 (Mendel)

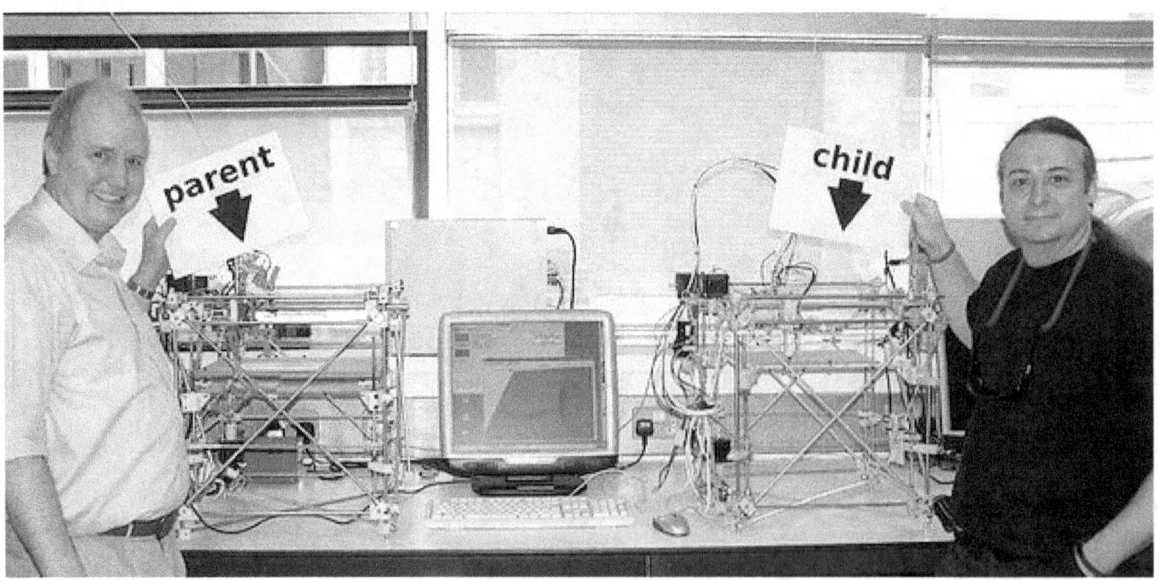

All of the plastic parts for the machine on the right were produced by the machine on the left. Adrian Bowyer (left) and Vik Olliver (right) are members of the RepRap project.

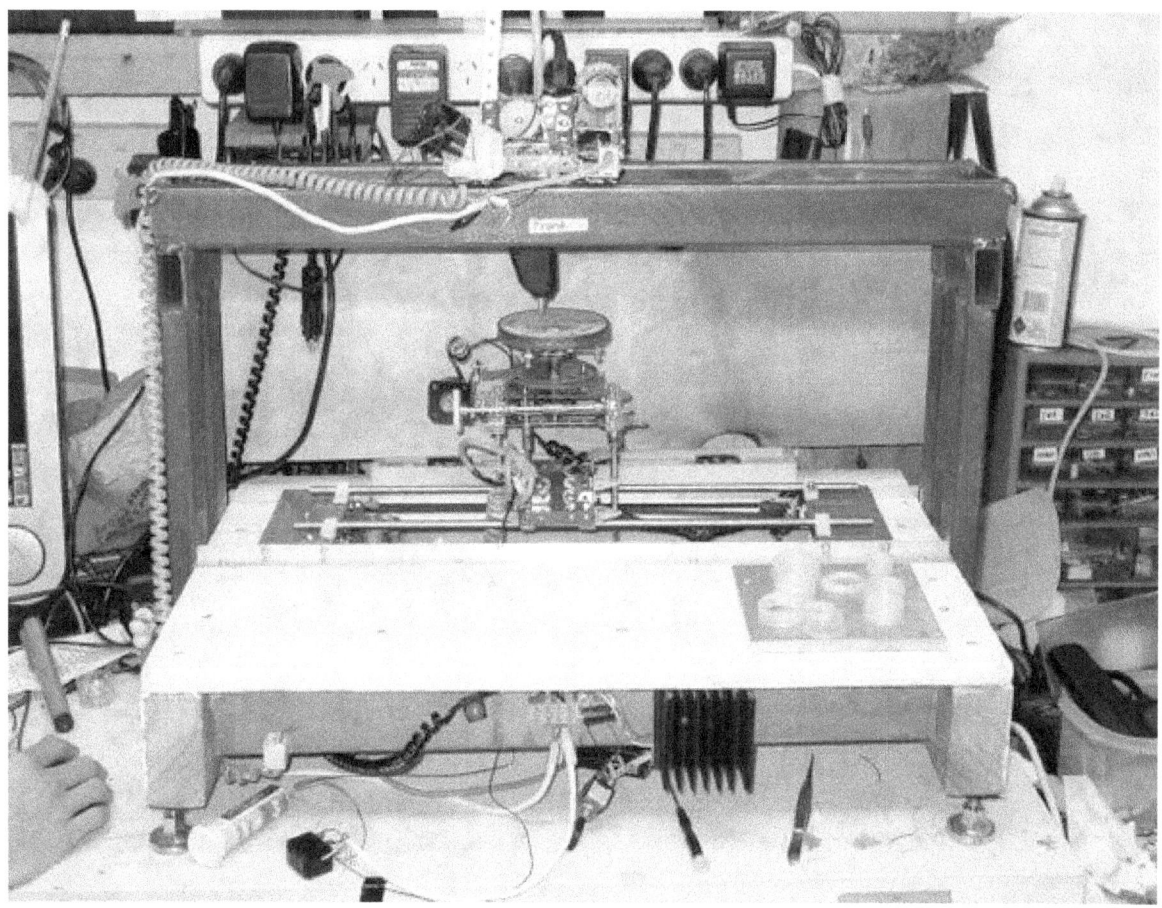

Meccano repstrap of RepRap 0.1 prototype (created by Vik Olliver)

The BigRep One.1 with its 1m³ volume.

Miniature face models (from FaceGen) produced using several colored plastics on a 3D Printer

The Audi RSQ was made with rapid prototyping industrial KUKA robots

An example of 3D printed limited edition jewellery. This necklace is made of glassfiber-filled dyed nylon. It has rotating linkages that were produced in the same manufacturing step as the other parts

Guardians of Time by Manfred Kielnhofer, 3D printing polished nickel steel by Shapeways 2014

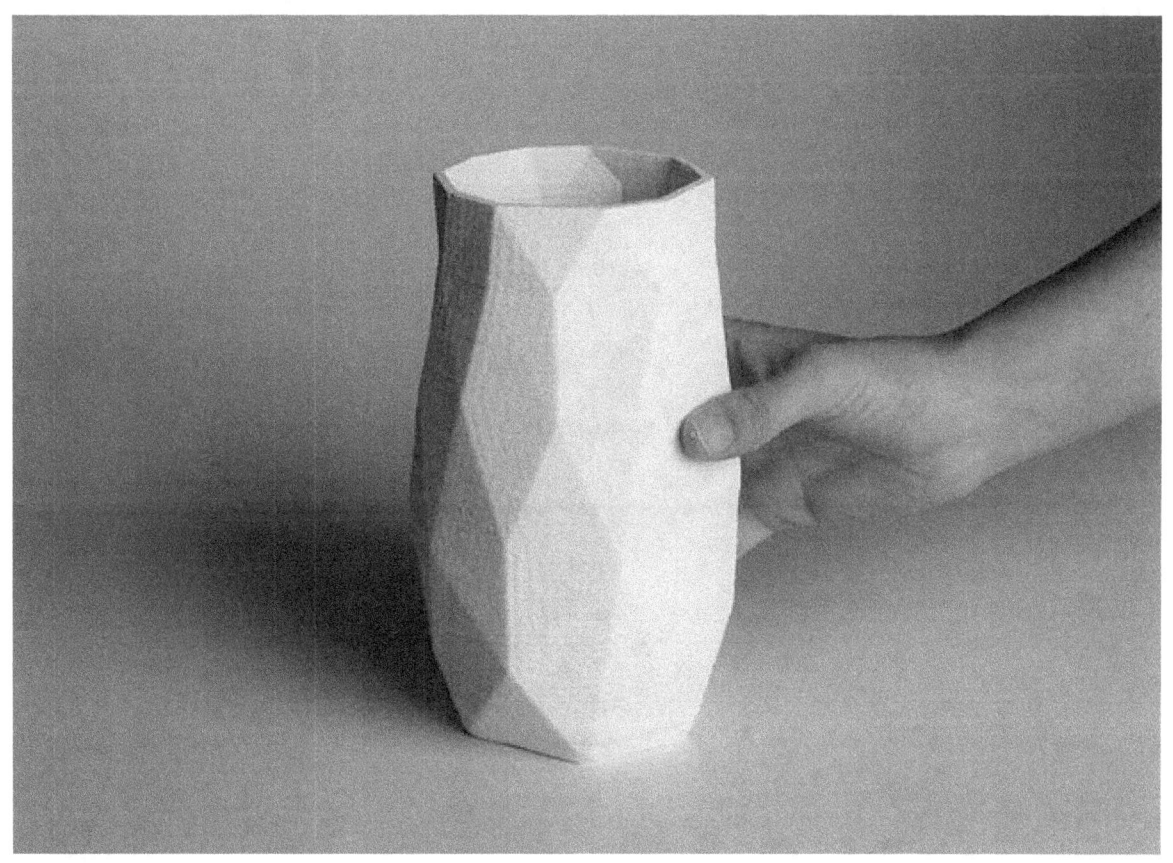

3D printed ceramic pot

flower model made with a 3D printer

Chapter 2

3D publishing

3D publishing concerns the production and distribution of content for 3D printers. 3D publishing holds out the promise of an industry for the creation and distribution of files for the production of 3D objects.

Any individual or organisation producing files for 3D printers can be considered a 3D publisher. With the advent of specialist software, scanners and cloud based tools, access to 3D publishing is spreading fast. The development of online tools to facilitate and monetize publishing is bringing a new industry to fruition. Boundaries between value chains are disappearing, leading to new business models. While 3D publishing and 3D publishers is a fairly new concept, a lot of development is happening in this space alongside the breakneck development of 3D printing hardware and software.

2.1 Business models

Free distribution Anyone can upload 3D models to a site and anybody can download the model and 3D print at home for free. Pinshape, Thingiverse, Youmagine, Clara.io, MyMiniFactory, Threeding.

Shopfront Shopfront services allow anyone to open a shop and upload their 3D models. Customers pay to get the 3D model printed via the 3D print services of these companies. Shapeways, Ponoko, i.Materialise, Sculpteo, MyMiniFactory, Threeding. The designers of the selected 3D models will get a fee.

Paid distribution Designers can upload designs and make them available for paid download for profit. Pinshape.

Hybrid In a hybrid model, the designer or company might make use of any of the above services and/or local 3D print bureau to create print to order models.

2.2 Ecosystem

There are many more 3D content strategies being developed as part of a growing industry of 3D printing. Companies that are not 3D print specialists are entering the field. These may be companies who have existing content and brands that can be distributed as 3D models. These companies need tools that enable the secure and efficient distribution of content. Employees will be required who are able to create, select, edit and market 3D content as part of the strategy of the company. In the near future, companies like IKEA may offer a database with accessories where the customer could pay, download and 3D print it, and use it on their IKEA table.

A conference on 3D publishing[1] was first held in the Netherlands in March 2013.

2.3 Software tools

In order to create 3D models one needs 3D software tools. There is a broad range of tools available from very simple tools on an tablet to very sophisticated engineering tools.

2.4 Examples

Hasbro, the brand behind My Little Pony and Transformers has joined with 3D Systems for a new 3D printing project.[2] Children will be able to 3D print their own toys.

A book called LEO the Maker Prince,[3] which is the First Children's Book On 3D Printing. The book was made in order to give young children a chance to get involved with this new way of producing objects.

Trobok Digital Toy Store[4] has opened their website with their own designed toys with fancy names like SPIKE, 21st Century Bear, Pig Corp! and SCOUT. The customer has to pay for the design, than download it and 3D print it at home.

Amazon has opened a 3D Printing Store[5] for the sale of models for 3D printing. Designers can use the store to sell printed models.

2.5 See also

- 3D modeling

- 3D scanner

- 3D Printing Marketplace

2.6 References

[1] "3D Publishing Conference". Jakajima. Retrieved 3 August 2014.

[2] "Hasbro".

[3] "Leo the Maker Prince".

[4] "Trobok".

[5] "Amazon 3D Print Store". *Amazon*. Retrieved 3 August 2014.

2.7 External links

- The Coming Ecosystem of 3D Printing

- Introduction to 3d printing costs

- Structural optimisation of metallic components

- 3D fabbers: don't let the DMCA stifle an innovative future // Arstechnica, 2010-11-10

- 3-D printing at MIT

- 3D Printing: The Printed World from *The Economist*

- 3D Printer News and Updates from Industry

- 3D Printing Industry News

- New 3D design and mind

- Overview of recent 3D printing applications (from Dezeen magazine)

- Will 3D Printing Change the World? Video produced by Off Book (web series)

Chapter 3

2BOT Physical Modeling Technologies

2BOT *physical* **Modeling Technologies** is a privately held company headquartered in Redmond, WA that develops, manufactures, and distributes 3D model making hardware and software.[1]
In June 2010, 2BOT launched their 3D model maker, the ModelMaker, and accompanying software, 2BOT Studio.[2]
Currently, the company offers its ModelMaker to classroom and professional settings.

3.1 References

[1] Rapid Today. "GIS 3D Printing Made Easier With Software". Archived from the original on 15 July 2011. Retrieved 2011-06-29.

[2] WebWire. "The 2Bot ModelMaker: A Game Changing Tool For Architects & GIS Professionals, to debut at the 2010 AIA Convention". Retrieved 2011-06-29.

3.2 External links

- 2BOT's Official Website

Chapter 4

3D bioprinting

3D bioprinting is the process of generating spatially-controlled cell patterns using 3D printing technologies, where cell function and viability are preserved within the printed construct.[1]:1 The first patent related to this technology was filed in the United States in 2003 and granted in 2006.[1]:1 [2]

4.1 Process

Using 3D bioprinting for fabricating biological constructs typically involves dispensing cells onto a biocompatible scaffold using a successive layer-by-layer approach to generate tissue-like three-dimensional structures.[3]:3 Artificial organs such as livers and kidneys made by 3D bioprinting have been shown to lack crucial elements that affect the body such as working blood vessels, tubules for collecting urine, and the growth of billions of cells required for these organs. Without these components the body has no way to get the essential nutrients and oxygen deep within their interiors.[3]:3 Given that every tissue in the body is naturally compartmentalized of different cell types, many technologies for printing these cells vary in their ability to ensure stability and viability of the cells during the manufacturing process. Some of the methods that are used for 3D bioprinting of cells are photolithography, magnetic bioprinting, stereolithography, and direct cell extrusion. Typically, the first step used is getting a biopsy of the organ. From this examination, certain cells are isolated and multiplied. These cells are then mixed with a special liquefied material that provides oxygen and other nutrients to keep them alive. Finally, the mixture is placed in a printer cartridge and structured using the patients' medical scans. [4]:4When a bioprinted pre-tissue is transferred to an incubator this cell-based pre-tissue matures into a tissue.

4.2 Applications

San Diego-based Organovo, an "early-stage regenerative medicine company", was the first company to commercialize 3D bioprinting technology.[1]:1 The company utilizes its NovoGen MMX Bioprinter for 3D bioprinting. The printer is optimized to be able to print skin tissue, heart tissue, and blood vessels among other basic tissues that could be suitable for surgical therapy and transplantation. A research team at Swansea University in the UK is using Bioprinting technology to produce soft tissues and artificial bones for eventual use in reconstructive surgery. Bioprinting technology will eventually be used to create fully functional human organs for transplants and drug research. This will allow for more effective organ transplants and safer more effective drugs.[5]:6

4.3 Impact

3D-bioprinting attributes to significant advances in the medical field of tissue engineering by allowing for research to be done on innovative materials called biomaterials. Biomaterials are the materials adapted and used for printing three-dimensional objects. Some of the most notable bioengineered substances that are usually stronger than the average bodily materials, including soft tissue and bone. These constituents can act as future substitutes, even improvements, for the original body materials. Alginate, for example, is an anionic polymer with many biomedical implications including feasibility, strong biocompatibility, low toxicity, and stronger structural ability in comparison to some of the body's structural material.[6]:7 Synthetic hydrogels are also commonplace, including PV based gels. The combination of Acid with a UV initiated PV based cross-linker has been evaluated by the Forest Institute of Medicine and determined to be a very biomaterial.[7] Engineers are also exploring other options such as printing micro-channels that can maximize the diffusion of nutrients and oxygen from neighboring tissues [4]:4 In addition, The Defense Threat Reduction Agency aims to print mini organs such as hearts, livers, and lungs as the potential to test new drugs more accurately and perhaps eliminate the need for testing in animals [4]:4

4.4   See also

- 3D bio-printing section in the 3D printing article

- Magnetic 3D Bioprinting

4.5   References

[1] Doyle, Ken (15 May 2014). "Bioprinting:From Patches to Parts". *Gen. Eng. Biotechnol. News* (paper) **34** (10): 1, 34–5. abstract

[2] US patent 7051654, Boland, Thomas; Wilson, Jr., William Crisp; Xu, Tao, "Ink-jet printing of viable cells", issued 2006-05-30

[3] Harmon K. A Sweet Solution for Replacing Organs. Sci Am 2013 −04-01;308(4):54-55.

[4] Cooper-White, M. "How 3D Printing Could End The Deadly Shortage Of Donor Organs." 2015.

[5] Thomas D. *Engineering Ourselves – The Future Potential Power of 3D-Bioprinting?* 2015; .

[6] Crawford M. *Creating Valve Tissue Using 3-D Bioprinting* 2015;

[7] Murphy, S. V.; Skardal, A; Atala, A (2013). "Evaluation of hydrogels for bio-printing applications". *Journal of Biomedical Materials Research Part A* **101** (1): 272–84. doi:10.1002/jbm.a.34326. PMID 22941807.

Chapter 5

3D Hubs

3D Hubs is an online 3D printing service platform. It operates a network of 3D printers with over 20,000 locations in over a 150 countries, providing over 1 billion people access to a 3D printer within 10 miles of their home.[1][2] The company facilitates transactions between 3D printer owners (Hubs) and people that want to make 3D prints.[3] Printer owners can join the platform to offer 3D printing services while customers can locate printer owners to get their 3D models printed nearby.

5.1   Company

3D Hubs facilitates transactions between 3D printer owners (Hubs) and people who want to make 3D prints. The company was founded in April 2013 by Bram de Zwart and Brian Garret. Headquartered in Amsterdam, the company opened its second office in New York in August, 2014.

3D Hubs is a privately held company backed by Balderton Capital and Dutch investors DOEN and Zeeburg. The company raised $4.5 million in its Series A round of funding in September, 2014.[4][5] As part of the financial deal, Mark Evans, General Partner at Balderton Capital, joined the two founders on the company board.

The company has formed partnerships with Autodesk and Fairphone and Uber. The recent partnership with Uber rush allows users to have prints shipped within NYC using Uber's rush service with discounted rates

5.2   Trend report

The company releases a monthly trend report on the state of the 3D printing industry, including print quality ratings, 3D printer model popularity, print categories, material and colour choices. The report is based on data from 12,000 3D printers and more than 30,000 orders of the 3D Hubs community.[6]

5.3   See also

- 3D Printing

- 3D Printing Marketplace

5.4 References

[1] Park, Rachel (3 September 2014). "A Cash Injection for 3D Hubs' 3D Printing Model Will Accelerate Growth Even More". 3D Printing Industry. Retrieved 16 October 2014.

[2] Sher, Davide (4 March 2015). "Vibrant 3D Hubs Community Grows by 20% in Post-Holiday Period". 3D Printing Industry. Retrieved 6 March 2015.

[3] Horn, Leslie (27 October 2013). "3D Hubs: Like Airbnb For 3D Printers". Gizmodo. Retrieved 16 October 2014.

[4] PitchBook Data Inc. (3 Feb 2015). "Is Europe at the heart of 3D printing innovation? Here's why it may be". Hot Topics. Retrieved 10 February 2015.

[5] Biggs, John (2 Sep 2014). "3DHubs Raises $4.5 Million To Make Local 3D Printing Global". TechCrunch. Retrieved 16 October 2014.

[6] Molitch-Hou, Michael (26 September 2014). "3D Hubs Tracks Big Crowdfunding Growth for 3D Printing Industry in October Trend Report". 3D Printing Industry. Retrieved 16 October 2014.

5.5 External links

- Official website

Chapter 6

3D Manufacturing Format

3D Manufacturing Format or **3MF** is File format publish by 3MF Consortium.[1]

3MF is an XML-based data format designed for using additive manufacturing, including information about materials, colors, and other information that cannot be represented in the STL (file format) format.[2]

As of today, CAD software related companies such as Autodesk, Dassault Systems and Netfabb are part of the 3MF Consortium. Other firms in the 3MF Consortium are Microsoft (for Operating System support), SLM and HP, whilst Shapeways are also included to give insight from a 3D Printing background.[3] Other key players in the 3D printing and additive manufacturing business, such as Materialise, 3D Systems, Siemens PLM Software and Stratasys have recently joined the consortium.[4]

6.1 References

[1] "3MF Website". *3MF*. Retrieved 1 May 2015.

[2] "What is 3MF?".

[3] "3MF Consortium Launches to Advance 3D Printing Technology". Business Wire. Retrieved 1 May 2015.

[4] "3MF Consortium Signs New Members 3D Systems, Materialise, Siemens PLM Software and Stratasys"

Chapter 7

3D Print Canal House

The **3D Print Canal House** is a three-year publicly accessible "Research & Design by Doing" project in which an international team of partners from various sectors works together on 3D printing a canal house in Amsterdam.[1]

By building the house, all parties research the possibilities of 3D printing architecture and form connections between design, science, culture, building, software, communities and the city. The project serves as both an exhibition of 3D printing technology, as well as a research site into 3D printing architecture. The project is initiated by DUS architects and the site, in Amsterdam North, opened to the public on March 1, 2014.[2][3]

7.1 Kamermaker

The house is being constructed by a machine called the KamerMaker ("RoomBuilder") which implies how the house will be built: room-by-room. The house is printed with the KamerMaker – a gigantic FDM printer that can print elements of 2 x 2 x 3,5 meters, developed by DUS. It is a movable pavilion which has the size of a shipping container. The machine can print components up to 2.2 by 2.2 by 3.5 meters in size. The machine itself is 6 meters tall. The Kamermaker can be moved by a truck or ship.[4]

7.2 References

[1] "First 3D Printed House to Be Built In Amsterdam". *ArchDaily*. Retrieved 22 August 2015.

[2] "OPP.Today - How 3D printing by robots is set to transform building". *OPP.Today*. Retrieved 22 August 2015.

[3] "The printed house, coming soon: Futurists see 3-D technology radically changing the way houses are built". *The Globe and Mail*. Retrieved 22 August 2015.

[4] "3D Print Canal House". *iamsterdam.com*. Retrieved 22 August 2015.

7.3 References

- Official website

Chapter 8

3D printed firearms

In 2012, the U.S.-based group Defense Distributed disclosed plans to design a working plastic gun that could be downloaded and reproduced by anybody with a 3D printer."[1][2] Defense Distributed has also designed a 3D printable AR-15 type rifle lower receiver (capable of lasting more than 650 rounds) and a variety of magazines, including ones for AK-47.[3] Soon after Defense Distributed succeeded in designing the first working blueprint to produce a plastic gun with a 3D printer in May 2013, the United States Department of State demanded that they remove the instructions from their website.[4]

In 2013 a Texas company, Solid Concepts, demonstrated a 3D printed version of an M1911 pistol made of metal, using an industrial 3D printer.[5]

8.1 Effect on gun control

After Defense Distributed released their plans, questions were raised regarding the effects that 3D printing and widespread consumer-level CNC machining[6][7] may have on gun control effectiveness.[8][9][10][11]

The U.S. Department of Homeland Security and the Joint Regional Intelligence Center released a memo stating that "*significant advances in three-dimensional (3D) printing capabilities, availability of free digital 3D printer files for firearms components, and difficulty regulating file sharing may present public safety risks from unqualified gun seekers who obtain or manufacture 3D printed guns,*" and that "*proposed legislation to ban 3D printing of weapons may deter, but cannot completely prevent their production. Even if the practice is prohibited by new legislation, online distribution of these digital files will be as difficult to control as any other illegally traded music, movie or software files.*"[12]

Internationally, where gun controls are generally tighter than in the United States, some commentators have said the impact may be more strongly felt, as alternative firearms are not as easily obtainable.[13] European officials have noted that producing a 3D printed gun would be illegal under their gun control laws,[14] and that criminals have access to other sources of weapons, but noted that as the technology improved the risks of an effect would increase.[15][16] Downloads of the plans from the UK, Germany, Spain, and Brazil were heavy.[17][18]

Attempting to restrict the distribution over the Internet of gun plans has been likened to the futility of preventing the widespread distribution of DeCSS which enabled DVD ripping.[19][20][21][22] After the US government had Defense Distributed take down the plans, they were still widely available via The Pirate Bay and other file sharing sites.[23] Some US legislators have proposed regulations on 3D printers, to prevent them being used for printing guns.[24][25] 3D printing advocates have suggested that such regulations would be futile, could cripple the 3D printing industry, and could infringe on free speech rights.[26][27][28][29][30][31][32]

8.2 Legal status

8.2.1 United States

In the United States, it is legal for individuals to manufacture firearms for personal use without a license, however this does not extend to some firearms such as Title II weapons (machine guns, suppressors etc.) or Assault Weapons in jurisdictions that still ban them.

Under the Undetectable Firearms Act any firearm that cannot be detected by a metal detector is illegal to manufacture, so legal designs for firearms such as the Liberator require a metal plate to be inserted into the printed body. The act had a sunset provision to expire December 9, 2013. Senator Charles Schumer proposed renewing the law, and expanding the type of guns that would be prohibited.[33] Proposed renewals and expansions of the current Undetectable Firearms Act (H.R. 1474, S. 1149) include provisions to criminalize individual production of firearm receivers and magazines that do not include arbitrary amounts of metal, measures outside the scope of the original UFA and not extended to cover commercial manufacture.[34][35] These "modernization" proposals have been criticized as disingenuous attempts to suppress adoption of and experimentation with 3D printers in home gunsmithing.[36]

On December 3, 2013, the United States House of Representatives passed the bill To extend the Undetectable Firearms Act of 1988 for 10 years (H.R. 3626; 113th Congress).[37] The bill extended the Act, but did not change any of the law's provisions.[38]

8.2.2 Japan

In Japan, in May 2014, Yoshitomo Imura was the first person to be arrested for possessing printed guns. Imura had five guns, two of which were capable of being fired, but had no bullets. Imura had previously posted blueprints and video of his guns to the Internet, which triggered the investigation.[39]

8.3 See also

- 3D Printing

- Defense Distributed

- Gun control

- Gun politics in the United States

- Improvised firearm

- List of notable 3D printed weapons and parts

8.4 References

[1] Greenberg, Andy (August 23, 2012). "'Wiki Weapon Project' Aims To Create A Gun Anyone Can 3D-Print At Home". *Forbes*. Retrieved August 27, 2012.

[2] Poeter, Damon (August 24, 2012). "Could a 'Printable Gun' Change the World?". *PC Magazine*. Retrieved August 27, 2012.

[3] Farivar, Cyrus (March 1, 2013). ""Download this gun": 3D-printed semi-automatic fires over 600 rounds". *Ars Technica*. Retrieved February 5, 2015.

[4] "Blueprints for 3-D printer gun pulled off website". www.statesman.com. Retrieved 2013-11-10.

[5] Gross, Doug (2013-11-09). "Texas company makes metal gun with 3-D printer". CNN. Retrieved 9 November 2013.

[6] "3D Printers, Meet Othermill: A CNC machine for your home office (VIDEO)". Guns.com. Retrieved 2013-11-10.

[7] Clark (6 October 2011). "The Third Wave, CNC, Stereolithography, and the end of gun control". PopeHat.com.

[8] Rosenwald, Michael S. (2013-02-25). "Weapons made with 3-D printers could test gun-control efforts". *Washington Post.*

[9] "Making guns at home: Ready, print, fire". The Economist. 2013-02-16. Retrieved 2013-11-10.

[10] Rayner, Alex (6 May 2013). "3D-printable guns are just the start, says Cody Wilson". *The Guardian* (London).

[11] Manjoo, Farhad (2013-05-08). "3-D-printed gun: Yes, it will be possible to make weapons with 3-D printers. No, that doesn't make gun control futile". Slate.com. Retrieved 2013-11-10.

[12] "Homeland Security bulletin warns 3D-printed guns may be 'impossible' to stop". Fox News. 2013-05-23. Retrieved 2013-11-10.

[13] Cochrane, Peter (2013-05-21). "Peter Cochrane's Blog: Beyond 3D Printed Guns". TechRepublic. Retrieved 2013-11-10.

[14] Gilani, Nadia (2013-05-06). "Gun factory fears as 3D blueprints put online by Defense Distributed | Metro News". Metro.co.uk. Retrieved 2013-11-10.

[15] "Liberator: First 3D-printed gun sparks gun control controversy". Digitaljournal.com. Retrieved 2013-11-10.

[16] "First 3D Printed Gun 'The Liberator' Successfully Fired - IBTimes UK". Ibtimes.co.uk. 2013-05-07. Retrieved 2013-11-10.

[17] "US demands removal of 3D printed gun blueprints". neurope.eu. Retrieved 2013-11-10.

[18] "España y EE.UU. lideran las descargas de los planos de la pistola de impresión casera | Economía | EL PAÍS". Economia.elpais.com. 2013-05-09. Retrieved 2013-11-10.

[19] "Controlled by Guns". Quiet Babylon. 2013-05-07. Retrieved 2013-11-10.

[20] "3dprinting | Jon Camfield dot com". Joncamfield.com. Retrieved 2013-11-10.

[21] "State Dept Censors 3D Gun Plans, Citing 'National Security' - News from Antiwar.com". News.antiwar.com. 2013-05-10. Retrieved 2013-11-10.

[22] "Wishful Thinking Is Control Freaks' Last Defense Against 3D-Printed Guns - Hit & Run". Reason.com. 2013-05-08. Retrieved 2013-11-10.

[23] "The Pirate Bay steps in to distribute 3-D gun designs". Salon.com. 2013-05-10. Retrieved 2013-11-10.

[24] "Sen. Leland Yee Proposes Regulating Guns From 3-D Printers « CBS Sacramento". Sacramento.cbslocal.com. 2013-05-08. Retrieved 2013-11-10.

[25] "Schumer Announces Support For Measure To Make 3D Printed Guns Illegal « CBS New York". Newyork.cbslocal.com. 2013-05-05. Retrieved 2013-11-10.

[26] "+ Downloads & Extras:". Makezine.com. 2011-06-30. Retrieved 2013-11-10.

[27] Ball, James (10 May 2013). "US government attempts to stifle 3D-printer gun designs will ultimately fail". *The Guardian* (London).

[28] "Like It Or Not, 3D Printing Will Probably Be Legislated". TechCrunch. 2013-01-18. Retrieved 2013-11-10.

[29] Liz Klimas (2013-02-19). "Engineer: Don't Regulate 3D Printed Guns, Regulate Explosive Gun Powder Instead | Video". TheBlaze.com. Retrieved 2013-11-10.

[30] Beckhusen, Robert (2013-02-15). "3-D Printing Pioneer Wants Government to Restrict Gunpowder, Not Printable Guns | Danger Room". Wired.com. Retrieved 2013-11-10.

[31] "How Defense Distributed Already Upended the World - Philip Bump". The Atlantic Wire. 2013-05-10. Retrieved 2013-11-10.

[32] Gayle S Putrich (13 May 2013). "Plastic gun draws eyes to 3-D printing". European Plastics News.

[33] "Senator seeks to extend ban on 'undetectable' 3D-printed guns". *the Guardian*. Retrieved 15 February 2015.

[34] H.R. 1474

[35] S. 1149

[36] "On Undetectable Firearms Act Renewal". *blog.defdist.org*. Defense Distributed. November 18, 2013. Retrieved 18 November 2013.

[37] "H.R. 3626 - All Actions". United States Congress. Retrieved 5 December 2013.

[38] "House votes to renew ban on plastic firearms". *Foxnews.com*. 3 December 2013. Retrieved 5 December 2013.

[39] "Japanese man arrested for possessing 3-D printer guns". Retrieved 15 February 2015.

8.5 External links

- How 3-D Printed Guns Evolved Into Serious Weapons in Just One Year, *Wired*, May 2014.

- Should We Be Afraid of the 3D Printed Gun?, *Popular Mechanics*, May 2014.

Chapter 9

3D printing marketplace

A **3D printing marketplace** is a website where users buy, sell and freely share digital 3D printable files for use on 3D printers. 3D printing marketplaces have emerged with the fast-growing segment of consumer 3D printers. Currently, the existing 3D printing marketplaces are handful and their business model is still not profitable.

9.1 Concept

The consumer market for 3D printers has grown tremendously over the past several years. According to Credit Suisse the growth in 2013 is 100% vs. 2012. Consumer 3D printers allow households to produce goods at home. Since most people are not CAD professionals, they have to use third party designs. 3D printing marketplaces are the largest sources of 3D printable designs and it is believed that they will dominate on the market of 3D printable objects.[1]

9.2 How 3D printing marketplaces work

3D printing marketplaces are a combination of file sharing websites, with or without a built in e-commerce capability. Designers upload suitable files for 3D printing whilst other users buy or freely download the uploaded files for printing. The marketplaces facilitate the account management, infrastructure, server resources and guarantees safe settlement of payments (e-commerce). Some of the marketplaces also offer additional services such as 3D printing on demand, location of commercial 3D print shops, associated software for model rendering and dynamic viewing of items using packages such as Sketchfab . The most widely used 3D printable file formats are stl, wrl and vrml.[2]

9.3 Type of 3D printing marketplaces

There are different varieties of 3D printing marketplaces. Some of them like Thingiverse are dedicated to free sharing of 3D printable files. Others, like Shapeways offer a 3D printing service for objects which have been provided for sale by designers. MyMiniFactory offers a combination of these two: their main activity being the free sharing or 3D printable files, they also offer print-on-demand and design-on-demand services. Another category are websites exemplified by Threeding and 3DPrintWise. These offer free and commercial exchange of digital 3D printable files for use on 3D printers but do not directly include 3D printing services themselves. These marketplaces do however, offer integration to databases of 3D printers provided by third parties such as MakeXYZ and 3D Hubs. These latter two resources each contain geo-location services to several thousands of registered 3D printers. The two largest personal 3D printers manufacturers Makerbot (part of Stratasys, Ltd) and Cubify (subsidiary of 3D Systems) offer their own file repositories for sharing, respectively Thingiverse and Cubify Store.[2] For professional 3D printing needs there are platforms which offer a reverse-bid style auction interface, an integrated escrow payment system and many features specifically tailored for B2B transactions.

9.4 Popular 3D printing marketplaces

Shapeways is a New York based 3D printing marketplace and an on-demand provider of 3D printing services. Designers upload design files, and users can place orders with Shapeways to produce the 3D printed item for them. Shapeways offers a variety of materials, including metals, plastics, ceramics, etc. The company houses 50 industrial printers and produces over a million products on demand.[3][4][5]

3DLT is a platform for 3D printing as-a-service through which retailers offer 3D printable products online and in-store.[6] Users of 3DLT design and upload 3D printable files and 3DLT works to print and sell these products.

Thingiverse offers free sharing of user-created digital designs for 3D printing. The website is owned by Makerbot (a subsidiary of Stratasys). Numerous technical projects use Thingiverse as a repository for shared innovation and dissemination of source materials to the public.[7]

3DShare is a website dedicated to the sharing of user-created digital design files. 3DShare allows users to share their designs, either for free or at the standard rate of 99c. By allowing users to share their files for free they want to form a legitimate sharing community of original designs. In order to incentivise designers to build a community that makes files available to all at an affordable cost, there is also the option to share files at the price of 99c.

MyMiniFactory offers free sharing of 3D printable files too, but unlike the other platforms, MyMiniFactory's content is fully curated,[8] meaning that every downloadable object has previously been tested on 3D printers. The website is property of iMakr and also offers a free streaming service[9] for 3D designers. They also provide print-on-demand and design-on-demand services.

Threeding is an Eastern European startup that offers free and paid 3D printable content. The company launched its services in 2013 and it became popular among CAD designers and hobbyists with its simple design and interface. Significant parts of the 3D objects available at Threeding.com are digital copies of historical artifacts.[10][11] Other sites have blossomed as market places for 3D printing such as Scultpteo.com and 3DPrintWise.com which have a commercial flavour. Both focussing on the explicit trading of files for 3D printing with integrations to 3rd party printers, modelling and rendering software.

Ponoko sells lasercutting designs as well as 3D printing designs, with a greater focus on the laser cutting side. It gained noticeable media attention because of its unique business model, as one of the first manufacturers that uses distributed manufacturing and on-demand manufacturing.

i.materialise offers a wide range of materials and especially finishes. It is the 3D printing marketplace of Belgian 3D printing company Materialise NV.

9.5 Copyright concerns

The current intellectual property legislation in the developed countries does not explicitly regulate 3D printing. This creates numerous questions about the statute of 3D printing marketplaces. Some analysts predict that 3D printing marketplaces will be "the next Napster". Most of the existing marketplaces are quite conservative on this topic. For example, Thingiverse removed the product Penrose triangle after a takedown notice submitted by the designer Ulrich Schwanitz. Most large 3D printing marketplaces also have procedures for copyright complaints. Further development of 3D printing and more new marketplaces for file sharing will most probably make copyright become a significant issue for them.[12][13]

9.6 See also

- AstroPrint

- 3D printing

- Clara.io

- MyMiniFactory

- Pinshape

- Sculpteo

- Sketchfab

- Shapeways

- Thingiverse

- Threeding

9.7   References

[1] "Credit Suisse: 3D printing market will be much bigger than what industry consultants estimate | 3D Printer News & 3D Printing News". 3ders.org. September 18, 2013.

[2] "Is there an "iTunes app store" for 3D printer models?". Quora. Retrieved January 12, 2014.

[3] Lovecraft, Raven (June 20, 2012). "Shapeways hits one million 3D printed creations". TG Daily. Retrieved January 12, 2014.

[4] "Shapeways | CrunchBase Profile". Crunchbase.com. Retrieved January 12, 2014.

[5] Sloan, Paul (April 23, 2013). "Shapeways, the Etsy of 3D printing, raises $30M | Cutting Edge – CNET News". News.cnet.com. Retrieved January 12, 2014.

[6] "3DLT Launches The First Store For Printable 3D Objects'", Techcrunch, April 30, 2013

[7] Baichtal, John (November 20, 2008). "Thingiverse.com Launches A Library of Printable Objects | GeekDad". Wired.com.

[8] *Make: 3D Printing: The Essential Guide to 3D Printers By Anna Kaziunas France*, Maker Media, Inc., 19 Nov 2013

[9] *3Dprint.com* MyMiniFactory TV is a Streaming Chance to Show Off Your 3D Design & Printing Skills

[10] "In Bulgaria, an eBay for 3D Printable Designs Called Threeding Emerges". On3dprinting.com. Retrieved January 12, 2014.

[11] 3dprintercafe.com on November 2, 2013 (December 31, 1969). "Threedingcom exchange of d printable files a8 : 3DPrintercafe | Latest news | 3D Printers | 3D Scanners | 3D Output technology". 3DPrintercafe. Retrieved January 12, 2014.

[12] "What's the Deal with Copyright and 3D Printing?". Public Knowledge. January 29, 2013. Retrieved January 12, 2014.

[13] "Can 3D printing avoid a Napster moment? — Tech News and Analysis". Gigaom.com. September 18, 2013.

Chapter 10

3D-printed spacecraft

3D printing began to be used in production versions of spaceflight hardware in early 2014. In January, SpaceX first flew a "Falcon 9 rocket with a 3D-printed Main Oxidizer Valve (MOV) body in one of the nine Merlin 1D engines". The valve is used to control flow of cryogenic liquid oxygen to the engine in a high-pressure, low-temperature, high-vibration physical environment.[1]

Other 3D-printed spacecraft assemblies have been ground-tested, but have not yet flown to space, including high-temperature, high-pressure rocket engine combustion chambers and the entire mechanical spaceframe and propellant tanks for a small satellite of a few hundred kilograms.

The new United Launch Alliance Vulcan launch vehicle—with first launch no earlier than 2019—is evaluating 3d-printing for over 150 parts: 100 polymer and 50+ metal parts.[2]

10.1 Applications

10.1.1 Rocket engines

The SuperDraco engine that provides launch escape system and propulsive-landing thrust for the Dragon V2 passenger-carrying space capsule is fully printed, and was the first fully printed rocket engine. In particular, the engine combustion chamber is printed of Inconel, an alloy of nickel and iron, using a process of direct metal laser sintering, and operates at a chamber pressure 6,900 kilopascals (1,000 psi) at a very high temperature. The engines are contained in a printed protective nacelle to prevent fault propagation in the event of an engine failure.[3][4][5] The SuperDracon engine produces 73 kilonewtons (16,400 lbf) of thrust.[6] The engine completed a full qualification test in May 2014, and is slated to make its first orbital spaceflight in 2015 or 2016.[1][5]

The ability to 3D print the complex parts was key to achieving the low-mass objective of the engine. It's a very complex engine, and it was very difficult to form all the cooling channels, the injector head, and the throttling mechanism. ... [The ability] "to print very high strength advanced alloys ... was crucial to being able to create the SuperDraco engine."[7]

In June 2014, Aerojet Rocketdyne (AJR) announced that they had "manufactured and successfully tested an engine which had been entirely 3D printed." The *Baby Banton* engine is a 22 kN (5,000 lbf) thrust engine that runs on LOX/kerosene propellant.[8] By March 2015, AJR had completed a series of hot-fire tests for additively manufactured components for its full-size AR-1 booster engine.[9]

10.1.2 Spacecraft structure

By 2014, 3D printing had begun to be used to print the entire mechanical structure and integral propellant tanks of a small spacecraft.[10]

59

3D-printed satellite mechanical structure, Arkyd-300, February 2014. The torus holds the propellant and provides the structural frame for the satellite.

10.2   References

[1] "SpaceX Launches 3D-Printed Part to Space, Creates Printed Engine Chamber for Crewed Spaceflight". SpaceX. Retrieved 2014-08-01. *Compared with a traditionally cast part, a printed valve body has superior strength, ductility, and fracture resistance, with a lower variability in materials properties. The MOV body was printed in less than two days, compared with a typical castings cycle measured in months. The valve's extensive test program – including a rigorous series of engine firings, component level qualification testing and materials testing – has since qualified the printed MOV body to fly interchangeably with cast parts on all Falcon 9 flights going forward.*

[2] Stone, Jeff (2015-04-21). "Vulcan Rocket: 3D Printing Launch Plan Includes More Than 100 Components". *International*

Business Times. Retrieved 22 April 2015.

[3] Norris, Guy (2014-05-30). "SpaceX Unveils 'Step Change' Dragon 'V2'". *Aviation Week*. Retrieved 2014-05-30.

[4] Kramer, Miriam (2014-05-30). "SpaceX Unveils Dragon V2 Spaceship, a Manned Space Taxi for Astronauts — Meet Dragon V2: SpaceX's Manned Space Taxi for Astronaut Trips". *space.com*. Retrieved 2014-05-30.

[5] Bergin, Chris (2014-05-30). "SpaceX lifts the lid on the Dragon V2 crew spacecraft". *NASAspaceflight.com*. Retrieved 2014-05-30.

[6] James, Michael; Salton, Alexandria; Downing, Micah (November 12, 2013), *Draft Environmental Assessment for Issuing an Experimental Permit to SpaceX for Operation of the Dragon Fly Vehicle at the McGregor Test Site, Texas, May 2014 – Appendices* (PDF), Blue Ridge Research and Consulting, LCC, p. 12, retrieved August 8, 2014

[7] Foust, Jeff (2014-05-30). "SpaceX unveils its "21st century spaceship"". *NewSpace Journal*. Retrieved 2014-05-31.

[8] "Aerojet Rocketdyne 3D Prints An Entire Engine in Just Three Parts". 3dprint.com. 2014-06-26. Retrieved 2014-08-08.

[9] "Aerojet Rocketdyne Hot-Fire Tests Additive Manufactured Components for the AR1 Engine to Maintain 2019 Delivery". Aerojet Rocketdyne. 2015-03-15. Retrieved 5 June 2015.

[10] Diamandis, Peter (2014-06-26). "Update from Planetary Resources". *Peter H. Diamandis channel*. Planetary Resources. Retrieved 2014-07-30.

Chapter 11

Additive Manufacturing File Format

For other uses of AMF, see AMF (disambiguation).

Additive Manufacturing File Format (**AMF**) is an open standard for describing objects for additive manufacturing processes such as 3D printing. The official ISO/ASTM 52915:2013[1]standard is an XML-based format designed to allow any computer-aided design software to describe the shape and composition of any 3D object to be fabricated on any 3D printer. Unlike its predecessor STL format, AMF has native support for color, materials, lattices, and constellations.

11.1 Structure

An AMF can represent one object, or multiple objects arranged in a constellation. Each object is described as a set of non-overlapping volumes. Each volume is described by a triangular mesh that references a set of points (vertices). These vertices can be shared among volumes. An AMF file can also specify the material and the color of each volume, as well as the color of each triangle in the mesh. The AMF file is compressed using the zip compression format, but the ".amf" file extension is retained. A minimal AMF reader implementation must be able to decompress an AMF file and import at least geometry information (ignoring curvature).

11.1.1 Basic file structure

The AMF file begins with the XML declaration line specifying the XML version and encoding. The remainder of the file is enclosed between an opening <amf> element and a closing </amf> element. The unit system can also be specified (millimeter, inch, feet, meter or micrometer). In absence of a units specification, millimeters are assumed.

Within the AMF brackets, there are five top level elements. Only a single object element is required for a fully functional AMF file.

1. **<object>** The object element defines a volume or volumes of material, each of which are associated with a material ID for printing. At least one object element must be present in the file. Additional objects are optional.

2. **<material>** The optional material element defines one or more materials for printing with an associated material ID. If no material element is included, a single default material is assumed.

3. **<texture>** The optional texture element defines one or more images or textures for color or texture mapping, each with an associated texture ID.

4. **<constellation>** The optional constellation element hierarchically combines objects and other constellations into a relative pattern for printing.

5. **<metadata>** The optional metadata element specifies additional information about the object(s) and elements contained in the file.

11.1.2 Geometry specification

The format uses a Face-vertex polygon mesh layout. The top level <object> element specifies a unique id. The <object> element can also optionally specify a material. The entire mesh geometry is contained in a single mesh element. The mesh is defined using one <vertices> element and one or more <volume> elements. The required <vertices> element lists all vertices that are used in this object. Each vertex is implicitly assigned a number in the order in which it was declared, starting at zero. The required child element <coordinates> gives the position of the point in 3D space using the <x>, <y> and <z> elements. After the vertex information, at least one <volume> element must be included. Each volume encapsulates a closed volume of the object, Multiple volumes can be specified in a single object. Volumes may share vertices at interfaces but may not have any overlapping volume. Within each volume, the child element <triangle> is used to define triangles that tessellate the surface of the volume. Each <triangle> element will list three vertices from the set of indices of the previously defined vertices given in the <vertices> element. The indices of the three vertices of the triangles are specified using the <v1>, <v2> and <v3> elements. The order of the vertices must be according to the right-hand rule, such that vertices are listed in counter-clockwise order as viewed from the outside. Each triangle is implicitly assigned a number in the order in which it was declared, starting at zero.

11.1.3 Color specification

Colors are introduced using the <color> element by specifying the red, green, blue and alpha (transparency) channels in the sRGB color space as numbers in the range of 0 to 1. The <color> element can be inserted at the material, object, volume, vertex, or triangle levels, and takes priority in reverse order (triangle color is highest priority). The transparency channel specifies to what degree the color from the lower level is blended in. By default, all values are set to zero.

A color can also be specified by referring to a formula that can use a variety of coordinate-dependent functions.

Texture maps

Texture maps allow assigning color or material to a surface or a volume, borrowing from the idea of Texture mapping in graphics. The <texture> element is first used to associate a texture-id with particular texture data. The data can be represented as either a 2D or a 3D array, depending on whether the color or material need to be mapped to a surface or a volume. The data is represented as a string of bytes in Base64 encoding, one byte per pixel specifying the grayscale level in the 0-255 range.

Once the texture-id is assigned, the texture data can be referenced in a color formula, such as in the example below.

Usually, however, the coordinated will not be used directly as shown above, but transformed first to bring them from object coordinates to texture coordinates. For example tex(1,f1(x,y,z),f2(x,y,z),f3(x,y,z)) where f1(), f2(), f3() are some functions, typically linear.

11.1.4 Material specification

Materials are introduced using the <material> element. Each material is assigned a unique id. Geometric volumes are associated with materials by specifying a material-id within the <volume> element.

Mixed, graded, lattice, and random materials

New materials can be defined as compositions of other materials. The element <composite> is used to specify the proportions of the composition, as a constant or as a formula dependent of the x, y, and z coordinates. A constant mixing proportion will lead to a homogenous material. A coordinate-dependent composition can lead to a graded material. More complex coordinate-dependent proportions can lead to nonlinear material gradients as well as periodic and non-periodic substructure. The proportion formula can also refer to a texture map using the tex(textureid,x,y,z) function. Reference to material-id "0" (void) is reserved and may be used to specify porous structures. Reference to the rand(x,y,z) function can be used to specify pseudo-random materials. The rand(x,y,z) function returns a random number between 0 and 1 that is persistent for that coordinate.

11.1.5   Print constellations

Multiple objects can be arranged together using the <constellation> element. A constellation can specify the position and orientation of objects to increase packing efficiency and to describe large arrays of identical objects. The <instance> element specifies the displacement and rotation an existing object needs to undergo to arrive into its position in the constellation. The displacement and rotation are always defined relatively to the original position and orientation in which the object was defined. A constellation can refer to another constellation as long as cyclic references are avoided.

If multiple top-level constellations are specified, or if multplie objects without constellations are specified, each of them will be imported with no relative position data. The importing program can then freely determine the relative positioning.

11.1.6   Meta-data

The <metadata> element can optionally be used to specify additional information about the objects, geometries and materials being defined. For example, this information can specify a name, textual description, authorship, copyright information and special instructions. The <metadata> element can be included at the top level to specify attributes of the entire file, or within objects, volumes and materials to specify attributes local to that entity.

11.1.7   Optional curved triangles

In order to improve geometric fidelity, the format allows curving the triangle patches. By default, all triangles are assumed to be flat and all triangle edges are assumed to be straight lines connecting their two vertices. However, curved triangles and curved edges can optionally be specified in order to reduce the number of mesh elements required to describe a curved surface. The curvature information has been shown to reduce the error of a spherical surface by a factor of 1000 as compared to a surface described by the same number of planar triangles.[1] Curvature should not create a deviation of from the plane of the flat triangle that exceeds 50% of the largest dimension of the triangle.

To specify curvature, a vertex can optionally contain a child element <normal> to specify desired surface normal at the location of the vertex. The normal should be unit length and pointing outwards. If this normal is specified, all triangle edges meeting at that vertex are curved so that they are perpendicular to that normal and in the plane defined by the normal and the original straight edge. When the curvature of a surface at a vertex is undefined (for example at a cusp, corner or edge), an <edge> element can be used to specify the curvature of a single non-linear edge joining two vertices. The curvature is specified using the tangent direction vectors at the beginning and end of that edge. The <edge> element will take precedence in case of a conflict with the curvature implied by a <normal> element.

When curvature is specified, the triangle is decomposed recursively into four sub-triangles. The recursion must be executed five levels deep, so that the original curved triangle is ultimately replaced by 1024 flat triangles. These 1024 triangles are generated "on the fly" and stored temporarily only while layers intersecting that triangle are being processed for manufacturing.

11.1.8   Formulas

In both the <color> and <composite> elements, coordinate-dependent formulas can be used instead of constants. These formulas can use various standard algebraic and mathematical operators and expressions.

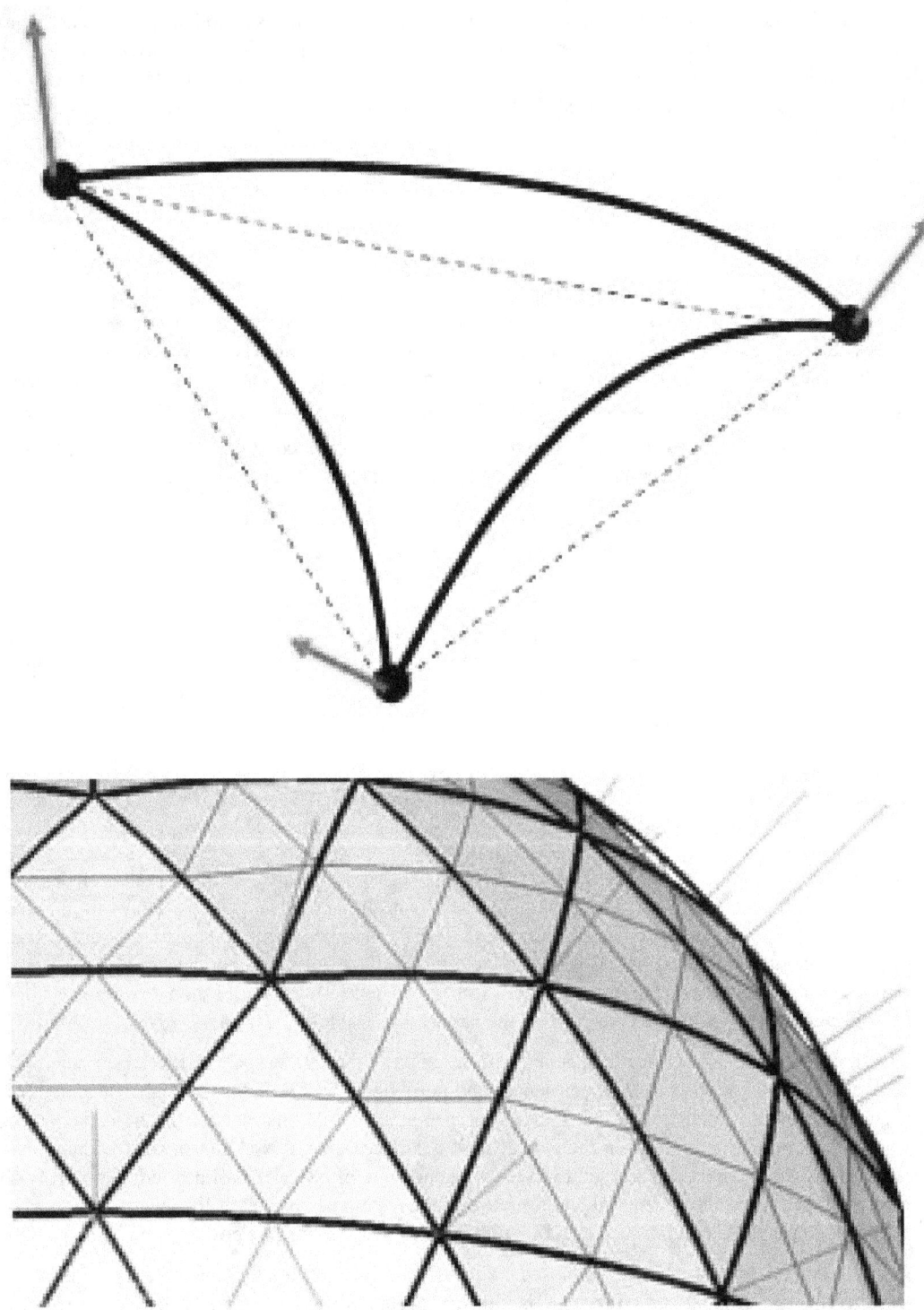

A curved triangle patch. Normals at vertices are used to recursively subdivide the triangle into four sub-triangles

11.1.9 Compression

An AMF can be stored either as plain text or as compressed text. If compressed, the compression is in ZIP archive format. A compressed AMF file is typically about half the size of an equivalent compressed binary STL file. The compression can be done manually using compression software such as WinZip, 7-Zip, or automatically by the exporting software during write. Both the compressed and uncompressed files have the AMF extension and it is the responsibility of the parsing program to determine whether or not the file is compressed, and if so to perform decompression during import.

11.2   Design considerations

When the ASTM Design subcommittee began developing the AMF specifications, a survey of stakeholders[2] revealed that the key priority for the new standard was the requirement for a non-proprietary format. Units and buildability issues were a concern lingering from problems with the STL format. Other key requirements were the ability to specify geometry with high fidelity and small file sizes, multiple materials, color, and microstructures. In order to be successful across the field of additive manufacturing, this file format was designed to address the following concerns

1. **Technology independence**: The file format must describe an object in a general way such that any machine can build it to the best of its ability. It is resolution and layer-thickness independent, and does not contain information specific to any one manufacturing process or technique. This does not negate the inclusion of properties that only certain advanced machines support (for example, color, multiple materials, etc.), but these are defined in such a way as to avoid exclusivity.

2. **Simplicity**: The file format must be easy to implement and understand. The format should be readable and editable in a simple text viewer, in order to encourage understanding and adoption. No identical information should be stored in multiple places.

3. **Scalability**: The file format should scale well with increase in part complexity and size, and with the improving resolution and accuracy of manufacturing equipment. This includes being able to handle large arrays of identical objects, complex repeated internal features (e.g. meshes), smooth curved surfaces with fine printing resolution, and multiple components arranged in an optimal packing for printing.

4. **Performance**: The file format must enable reasonable duration (interactive time) for read and write operations and reasonable file sizes for a typical large object.

5. **Backwards compatibility**: Any existing STL file should be convertible directly into a valid AMF file without any loss of information and without requiring any additional information. AMF files are also easily convertible back to STL for use on legacy systems, although advanced features will be lost.

6. **Future compatibility**: In order to remain useful in a rapidly changing industry, this file format must be easily extensible while remaining compatible with earlier versions and technologies. This allows new features to be added as advances in technology warrant, while still working flawlessly for simple homogenous geometries on the oldest hardware.

11.3   History

Since the mid-1980s, the STL file format has been the *de facto* industry standard for transferring information between design programs and additive manufacturing equipment. The STL format only contained information about a surface mesh, and had no provisions for representing color, texture, material, substructure, and other properties of the fabricated target object. As additive manufacturing technology evolved from producing primarily single-material, homogenous shapes to producing multi-material geometries in full color with functionally graded materials and microstructures, there was a growing need for a standard interchange file format that could support these features. A second factor that ushered the development of the standard was the improving resolution of additive manufacturing technologies. As the fidelity of printing processes approached micron scale resolution, the number of triangles required to describe smooth curved surfaces resulted in unacceptably large file sizes.

During the 1990s and 2000s, a number of proprietary file formats have been in use by various companies to support specific features of their manufacturing equipment, but the lack of an industry-wide agreement prevented widespread adoption of any single format. In January 2009, a new ASTM Committee F42 on Additive Manufacturing Technologies was established, and a design subcommittee was formed to develop a new standard. A survey was conducted in late 2009[2] leading to over a year of deliberations on the new standard. The resulting first revision of the AMF standard became official on May 2, 2011[3]

During the July 2013 meetings of ASTM's F42 and ISO's TC261 in Nottingham (UK), the Joint Plan for Additive Manufacturing Standards Development was approved. Since then, the AMF standard is managed jointly by ISO and ASTM.

11.4 Sample file

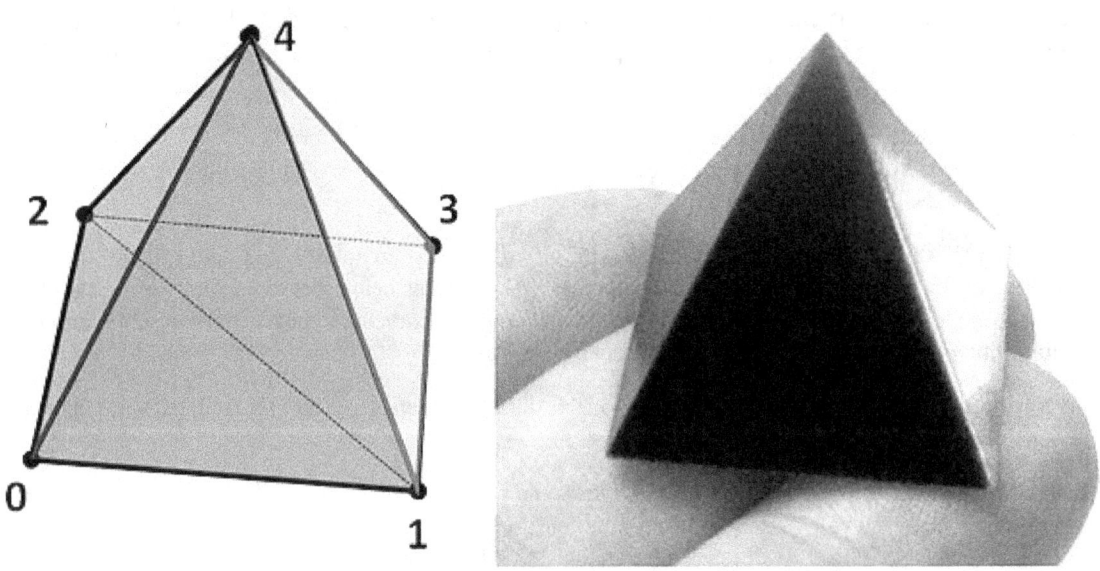

Object produced by the sample AMF code.

Below is a simple AMF file describing a pyramid made of two materials, adapted from the AMF tutorial[4] (548 bytes compressed). To create this AMF file, copy and paste the text below text into a text editor or an xml editor, and save the file as "pyramid.amf". Then compress the file with ZIP, and rename the file extension from ".zip" to ".zip.amf".

```
<?xml version="1.0" encoding="utf-8"?> <amf unit="inch" version="1.1"> <metadata type="name">Split Pyramid</metadata>
<metadata type="author">John Smith</metadata> <object id="1"> <mesh> <vertices> <vertex><coordinates><x>0</x><y>0</y><
<vertex><coordinates><x>1</x><y>0</y><z>0</z></coordinates></vertex> <vertex><coordinates><x>0</x><y>1</y><z>0</z
<vertex><coordinates><x>1</x><y>1</y><z>0</z></coordinates></vertex> <vertex><coordinates><x>0.5</x><y>0.5</y><z>1
</vertices> <volume materialid="2"> <metadata type="name">Hard side</metadata> <triangle><v1>2</v1><v2>1</v2><v3>0</v
<triangle><v1>0</v1><v2>1</v2><v3>4</v3></triangle> <triangle><v1>4</v1><v2>1</v2><v3>2</v3></triangle>
<triangle><v1>0</v1><v2>4</v2><v3>2</v3></triangle> </volume> <volume materialid="3"> <metadata type="name">Soft
side</metadata> <triangle><v1>2</v1><v2>3</v2><v3>1</v3></triangle> <triangle><v1>1</v1><v2>3</v2><v3>4</v3></tria
<triangle><v1>4</v1><v2>3</v2><v3>2</v3></triangle> <triangle><v1>4</v1><v2>2</v2><v3>1</v3></triangle>
</volume> </mesh> </object> <material id="2"> <metadata type="name">Hard material</metadata> <color><r>0.1</r><g>0.1</g
</material> <material id="3"> <metadata type="name">Soft material</metadata> <color><r>0</r><g>0.9</g><b>0.9</b><a>0.5
</material> </amf>
```

11.5   See also

- STL (file format)

- X3D

- 3D Printing Marketplace

- 3D Manufacturing Format

11.6   Notes

[1] Specification for Data Exchange Format for Additive Manufacturing

[2] STL 2.0 May Replace Old, Limited File Format Rapid Today, Oct 2009

[3] New ASTM Additive Manufacturing Specification Answers Need for Standard Interchange File Format ASTM, July 20, 2011

[4] AMF Tutorial: The Basics (Part 1)

11.7   External links

- AMF Wiki: A repository of AMF resources, sample files, and source code

Chapter 12

Alumide

Alumide is a material used in 3D printing consisting of nylon filled with aluminum dust, its name being a combination of the words aluminum and polyamide. Models are printed by sintering a tray of powder, layer by layer.[1] While it is much stiffer than other materials used in 3D printing, it can also withstand much higher thermal loads, maintaining its shape at temperatures that would cause thermoplastic compounds such as polylactic acid to become molten.[2]

[1] "Alumide". Shapeways. Retrieved 4 September 2013.

[2] "Materials and Material Management". Electro Optical Systems. Retrieved 4 September 2013.

Chapter 13

Aluminum polymer composite

An **aluminum polymer composite** (APC) material combines aluminum with a polymer to create materials with interesting characteristics. In 2014 researchers used a 3d laser printer to produce a polymer matrix. When coated with a 50-100 nanometer layer of aluminum oxide, the material was able to withstand loads of as much as 280 megapascals, stronger than any other known material whose density was less than 1,000 kilograms per cubic metre (1,700 lb/cu yd), that of water.[1][2]

13.1 Aluminum foam

Spherical aluminum foam pieces bonded by polymers produced foams that were 80-95% metal. Such foams were test=manufactured on an automated assembly line and are under consideration as automobile parts.

13.2 Thermal conductivity

Experimentally determined thermal conductivity of specific APCs matched both the Agari and Bruggeman models provide a good estimation for thermal conductivity. The experimental values of both thermal conductivity and diffusivity have shown a better heat transport for the composite filled with large particles.[3]

13.3 See also

- Aluminum foam

13.4 References

[1] Rathi, Akshat (2014-02-03). "New laser-printed material is lighter than water, as strong as steel". Ars Technica. Retrieved 2014-02-14.

[2] Bauer, J.; Hengsbach, S.; Tesari, I.; Schwaiger, R.; Kraft, O. (2014). "High-strength cellular ceramic composites with 3D microarchitecture". *Proceedings of the National Academy of Sciences*. doi:10.1073/pnas.1315147111.

[3] Boudenne, A.; Ibos, L.; Fois, M.; Gehin, E.; Majeste, J. C. (2004). "Thermophysical properties of polypropylene/aluminum composites". *Journal of Polymer Science Part B: Polymer Physics* **42** (4): 722. doi:10.1002/polb.10713.

13.5 External links

- Clark, Joel P. (October 1998). "Future of Automotive Body Materials: Steel, Aluminum & Polymer Composites" (PDF). Retrieved February 2014.

Chapter 14

Arthur Mamou-Mani

Arthur Georges Joel Mamou-Mani (born 5 February 1983, Paris) is a French architect. Mamou-Mani is director of the architecture and design practice Mamou-Mani Ltd which specializes in a new kind of pop-up, digital fabrication led architecture.[1]

14.1 Biography

He is a lecturer at the University of Westminster[2] in London and owns a digital fabrication laboratory called the FabPub. Mamou-Mani has given speeches including the TEDx conference in the United States,[3] the Develop3D Live[4] Conference and the Taipei Technical University in Taiwan. His work was featured at the Process Exhibition in Shanghai[5] and at the Sto Werkstatt in London.[6] He currently lives in London.[7]

He studied at the École nationale supérieure d'architecture de Paris-Malaquais[8] and in London, in 2003, at the Architectural Association School of Architecture. He then worked at Zaha Hadid Architects, Ateliers Jean Nouvel and Proctor and Matthews Architects[9] for three years. In 2011, he started teaching Diploma Studio 10 at the university of Westminster with Toby Burgess. To allow their students to share their ideas, they both created the online platform WeWantToLearn.net[10] receiving 600,000 views since its creation.[11][12] Arthur also founded his practice Mamou-Mani ltd in 2011. The projects include *the Magic Garden* for Karen Millen[13][14][15] and the 3D Pop-Up Studio for the Xintiandi shopping mall in Shanghai, one of the first component-based, fully 3D Printed pavilion (with Andrei Jipa and Stephany Xu)[16][17] Another pop-up project is "The Fitting Room" designed in collaboration with James K. Cheung of ARUP Associates[18] a large origami tree made of 500 laser-cut polypropylene folded pieces.[19]

14.2 Awards

- 2013: Crown Estate's best RIBA display for "The Magic Garden" at Karen Millen's flagship store on Regent Street[20]

- 2014: VM & Display best Christmas display[21]

14.3 Gallery

- "The Magic Garden" for Karen Millen by Arthur Mamou-Mani

- Xintiandi 3D Printing Pop-Up Studio

- A Giant Laser-Cut Origami Tree at Xintiandi Shanghai

- The Wooden Waves at BuroHappoldEngineering

14.4 External links

- Official website
- Strategies Using Grasshopper®
- Interview, in the Shanghai Daily
- Abitare sull'acqua, in La Repubblica
- The future of Britain? floating cities and high-rise farmsy, in The Telgraph
- Stackable housing pods, underwater cities, printed houses: Looking ahead at the city of the future, in WTTV
- From floating cities to high rise farms: Experts outline the future of Britain's architecture, in The Independent

14.5 See also

- 3D printing
- List of TED speakers

14.6 References

[1] "Architect describes his use of digital fabrication". *shanghaidaily.com*.

[2] BURNING MAN FESTIVAL 2013, Ramboll

[3] "Watch "The architecture of joy: Arthur Mamou-Mani and Toby Burgess at TEDxBlackRockCity" Video at TEDxTalks". *ted.com*.

[4] "Speakers". *develop3dlive.com*.

[5] "Arthur Mamou-Mani". *process-exhibition.eu*.

[6] "Staring at the Sun". *sto.com*.

[7] Official website

[8] Rencontre avec Arthur Mamou-Mani, architecte, Coté Maison

[9] Chester 'super' zoo plans approved, Architects Journal

[10] "WeWantToLearn.net". *WeWantToLearn.net*.

[11] "WeWantToLearn.net". *WeWantToLearn.net*.

[12] Showtime: the Three Cubes Colliding kite, Wired

[13] The Fashion Audit: Fuelled up flicks / One-off wonders / Designer displays, The Independent

[14] The Magic Garden, Dezeen

[15] Attention to retail..., World Architecture News

[16] "3DP Store for Shanghai Fashion Week - 3D Printing Industry". *3D Printing Industry*.

[17] Hypecask 3D Printers official Sponsors of the project, Hypecask

[18] The Folded Tree with, ARUP Associates

[19] Dawn of the (Retail) Soul: Convergence Theory and the Language of Design at XinTianDi, Forbes

[20] BT Fresca Limited. "Karen Millen Celebrates British Design With Arthur Mamou-Mani SS13 RIBA Regent Street Windows - Karen Millen". *karenmillen.com.*

[21] 2014 Awards Winners, VM & Display Directory

Chapter 15

Building printing

Building printing refers to various technology that use 3D printing as a way to construct buildings. Potential advantages of this process include quicker construction, lower labor costs, and less waste produced. 3D printing at a large scale may be well suited for construction of extraterrestrial structures on the Moon or other planets where environmental conditions are less conducive to human labor-intensive building practices.

Developments in additive manufacturing technologies have included attempts to make 3D printers capable of producing structural buildings.

15.1 History

Related technology development began in the 1960s, with pumped concrete and isocyanate foams.[1]

15.2 Current technology

Modern development and research has been under way since 2004 to flexibly construct buildings for commercial and private habitation. With built-in plumbing and electrical facilities, in one continuous build the process uses large 3D printers that would notionally complete the building in approximately 20 hours of "printer" time.[2] By January 2013, working versions of 3D-printing building technology were printing 2 metres (6 ft 7 in) of building material per hour, with a follow-on generation of printers proposed to be capable of 3.5 metres (11 ft) per hour, sufficient to complete a building in a week.[3]

Behrokh Khoshnevis founded the Contour Crafting project which demonstrated the basic capability, based on two parallel rails, an XY-controlled printing gantry and pressurized concrete tank. Dutch architect Janjaap Ruijssenaars's performative architecture 3D-printed building was planned to be built by a partnership of Dutch companies.[4] [5] The house was planned to be build in the end of 2014, but this deadline wasn't met. The companies said that they are still 100% sure the house will be printed.[6]

Various approaches to building printing are being researched. Two of these are Contour crafting[7] and D-Shape.[8][9] Other approaches involve direct sintering of inorganic raw materials to build composite ceramic building structures, similar to the approach used with metals in direct metal laser sintering.[10]

15.2.1 3D printed residential buildings

In the Netherlands, DUS Architects is 3D printing a 3D Printed Canal House, together with a international team of partners. The 3D Print Canal House links science, design, construction and community at a open building site in the heart of Amsterdam. Their aim is to demonstrate how 3D printing could revolutionize construction by increasing efficiency and reducing pollution and waste, and offer new tailor made housing solutions worldwide. 3D printing could also play a significant role in the quick build of low-cost housing in impoverished areas and those affected by disasters. The 3D Print Canal House is currently under construction at a canal-side plot in Amsterdam – an open 'expo-site' that it is proving to be a popular visitor attraction for the public. At the heart of the site, is the Kamermaker, or Room Builder – which is essentially a scaled up version of a table-top 3D printer. The Kamermaker prints building blocks from molten bio-plastic. This is currently a mix of 80% plant oil reinforced with microfibers, although this formula is still under development with the project's materials partner Henkel. For reinforcement, the blocks have an internal honeycombed centre that can be back-filled with Eco concrete. It also provides space for pipes, wiring and data cables to be installed internally.

The building blocks are then used to form component parts that can be slotted together like Lego to create a 4-storey, 13-room structure modelled on a traditional Dutch canal house. One of the most distinct design features of the Canal House is its geometrically-faceted plastic façade. 3D Print House Building BlocksThis gives a contemporary 3D print twist to the traditional canal house silhouette. The ability to print ornamental detailing on demand is a key design benefit of 3D modelling and printing in the building industry. With costly labour-intensive work reduced, custom-designed homes would become more accessible. So what are the main benefits of printing a house? Waste materials are a big problem for the building industry, but with 3D printing only the necessary raw materials are produced for each project. An added bonus is that 3D printer 'ink' can be made from recycled plastic waste. If printing on site, transport costs and CO_2 emissions are greatly reduced – as are dust and noise levels. And when the building is no longer needed, it can be shredded and recycled. Another key driver for developing this technology within the construction industry is the growing need for rapidly-produced housing. In this respect, 3D printing has the potential to reshape the way in which we build our cities – especially as Megacities are on the increase around the globe. The 3D Print Canal House was the first full-scale construction project of its kind to get off the ground. In just a short space of time, the Kamermaker has been further developed to increase its production speed by 300%. However, progress has not been swift enough to claim the title of 'World's First 3D Printed House'. [11]

The Chinese company WinSun has built several houses using large 3D printers sparing a mixture of quick drying cement and recycled raw materials.[12] Ten demo houses were built in 24 hours, each costing US$5000.[13][14]

Dutch and Chinese demonstration projects are slowly constructing 3D-printed buildings, using the effort to educate the public to the possibilities of the new plant-based building technology and to spur greater innovation in 3D printing of residential buildings.[15][16]

15.2.2   Extraterrestrial printed structures

The printing of buildings has been proposed as a particularly useful technology for constructing off-Earth habitats, such as habitats on the Moon or Mars.

As of 2013, the European Space Agency was working with London-based Foster + Partners to examine the potential of printing lunar bases using regular 3D printing technology.[17] The architectural firm proposed a building-construction 3D-printer technology in January 2013 that would use lunar regolith raw materials to produce lunar building structures while using enclosed inflatable habitats for housing the human occupants inside the hardshell printed lunar structures. Overall, these habitats would require only ten percent of the structure mass to be transported from Earth, while using local lunar materials for the other 90 percent of the structure mass.[3]

The dome-shaped structures would be a weight-bearing catenary form, with structural support provided by a closed-cell structure, reminiscent of bird bones.[18] In this conception, "printed" lunar soil will provide both "radiation and temperature insulation" for the Lunar occupants.[3] The building technology mixes lunar material with magnesium oxide which will turn the "moonstuff into a pulp that can be sprayed to form the block" when a binding salt is applied that "converts [this] material into a stone-like solid."[3] A type of sulfur concrete is also envisioned.[18]

Tests of 3D printing of an architectural structure with simulated lunar material have been completed, using a large vacuum chamber in a terrestrial lab.[19] The technique involves injecting the binding liquid under the surface of the regolith with a 3D printer nozzle, which in tests trapped 2 millimetres (0.079 in)-scale droplets under the surface via capillary forces.[18] The printer used was the D-shape.

A variety of lunar infrastructure elements have been conceived for 3D structural printing, including landing pads, blast protection walls, roads, hangars and fuel storage.[18]

In early 2014, NASA funded a small study at the University of Southern California to further develop the *Contour Crafting* 3D printing technique. Potential applications of this technology include constructing lunar structures of a material that could consist of up to 90-percent lunar material with only ten percent of the material requiring transport from Earth.[7]

NASA is also looking at a different technique that would involve the sintering of lunar dust using low-power (1500 watt) microwave energy. The lunar material would be bound by heating to 1,200 to 1,500 °C (2,190 to 2,730 °F), somewhat below the melting point, in order to fuse the nanoparticle dust into a solid block that is ceramic-like, and would not require the transport of a binder material from Earth as required by the Foster+Partners, Contour Crafting, and D-shape approaches to extraterrestrial building printing. One specific proposed plan for building a lunar base using this technique would be called SinterHab, and would utilize the JPL six-legged ATHLETE robot to autonomously or telerobotically build lunar structures.[10]

15.3 See also

- Building construction

- Space habitat

- Made in Space[20]

15.4 References

[1] Papanek (1971). *Design for the Real World*. ISBN 978-0897331531.

[2] "3D printer can build a house in 20 hours". YouTube. 2012-08-13. Retrieved 2014-03-13.

[3] Diaz, Jesus (2013-01-31). "This Is What the First Lunar Base Could Really Look Like". *Gizmodo*. Retrieved 2013-02-01.

[4] 3D printed Landscape House

[5] "The World's First 3D-Printed Building Will Arrive In 2014". *TechCrunch*. 2012-01-20. Retrieved 2013-02-08.

[6] Video summary of Landscape house forum and workshop Sept 3rd 2014

[7] "NASA's plan to build homes on the Moon: Space agency backs 3D print technology which could build base". *TechFlesh*. 2014-01-15. Retrieved 2014-01-16.

[8] Edwards, Lin (19 April 2010). "3D printer could build moon bases". *Phys.org*. Retrieved 21 October 2013.

[9] Cesaretti, Giovanni; Enrico Dini; Xavier de Kestelier; Valentina Colla; Laurent Pambaguian (January 2014). "Building components for an outpost on the Lunar soil by means of a novel 3D printing technology". *Acta Astronautica* **93**: 430–450. doi:10.1016/j.actaastro.2013.07.034. Retrieved 4 November 2013.

[10] Steadman, Ian. "Giant Nasa spider robots could 3D print lunar base using microwaves (Wired UK)". Wired.co.uk. Retrieved 2014-03-13.

[11] http://rhinecapital.com/investmentinsights/3d-print-canal-house/

[12] "China's Building 3D Printed Houses". Investing.com. Retrieved 2014-08-23.

[13] "China: Firm 3D prints 10 full-sized houses in a day". www.bbc.com. Retrieved 2014-04-28.

[14] Giant 3D printer creates 10 full-sized houses in a DAY: Bungalows built from layers of waste materials cost less than £3,000 each, Daily Mail, 28 April 2014, accessed 16 May 2014.

[15] "How Dutch team is 3D-printing a full-sized house". BBC. 2014-05-03. Retrieved 2014-06-10.

[16] The plan to print actual houses shows off the best and worst of 3D printing (2014-06-26), James Robinson, *PandoDaily*

[17] "Building a lunar base with 3D printing / Technology / Our Activities / ESA". Esa.int. 2013-01-31. Retrieved 2014-03-13.

[18] "3D Printing of a lunar base using lunar soil will print buildings 3.5 meters per hour". *Newt Big Future*. 2013-09-19. Retrieved 2013-09-23.

[19] "3D printed moon building designs revealed". *BBC News*. 2013-02-01. Retrieved 2013-02-08.

[20] "NASA - 3D Printing In Zero-G Technology Demonstration". Nasa.gov. 2014-03-04. Retrieved 2014-03-13.

15.5   External links

- Contour Crafting Project from USC, 2004

- Future of Construction Process: 3D Concrete Printing, 2010.

- Lunar Base Using 3D Printing, video, 2013.

- 3D Printing of a lunar base using lunar soil will print buildings at 3.5 meters per hour, NextBigFuture, 2013

Chapter 16

CandyFab

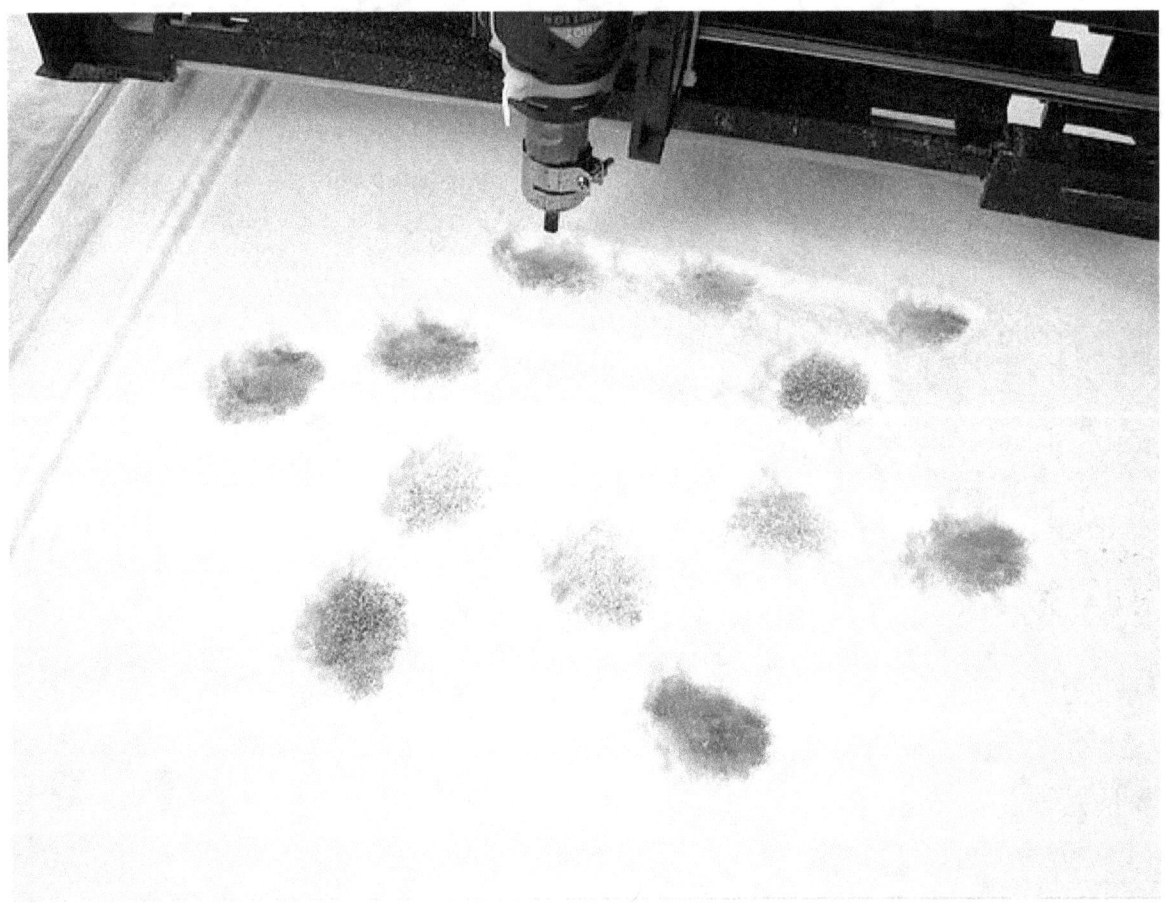

Midpoint in printing a toroidal coil sculpture on the CandyFab 4000, an experimental rapid prototyping machine that can use granulated sugar as the printing medium.

The **CandyFab** is a method of producing physical objects out of a computer representation of the structure. It differs from some other 3D printing methods in the following aspects:

- It is optimized for relatively large pieces using low to medium print resolution.

- To reduce the hazards of working with large amounts of media, non-toxic materials (ideally, food-grade) are used.

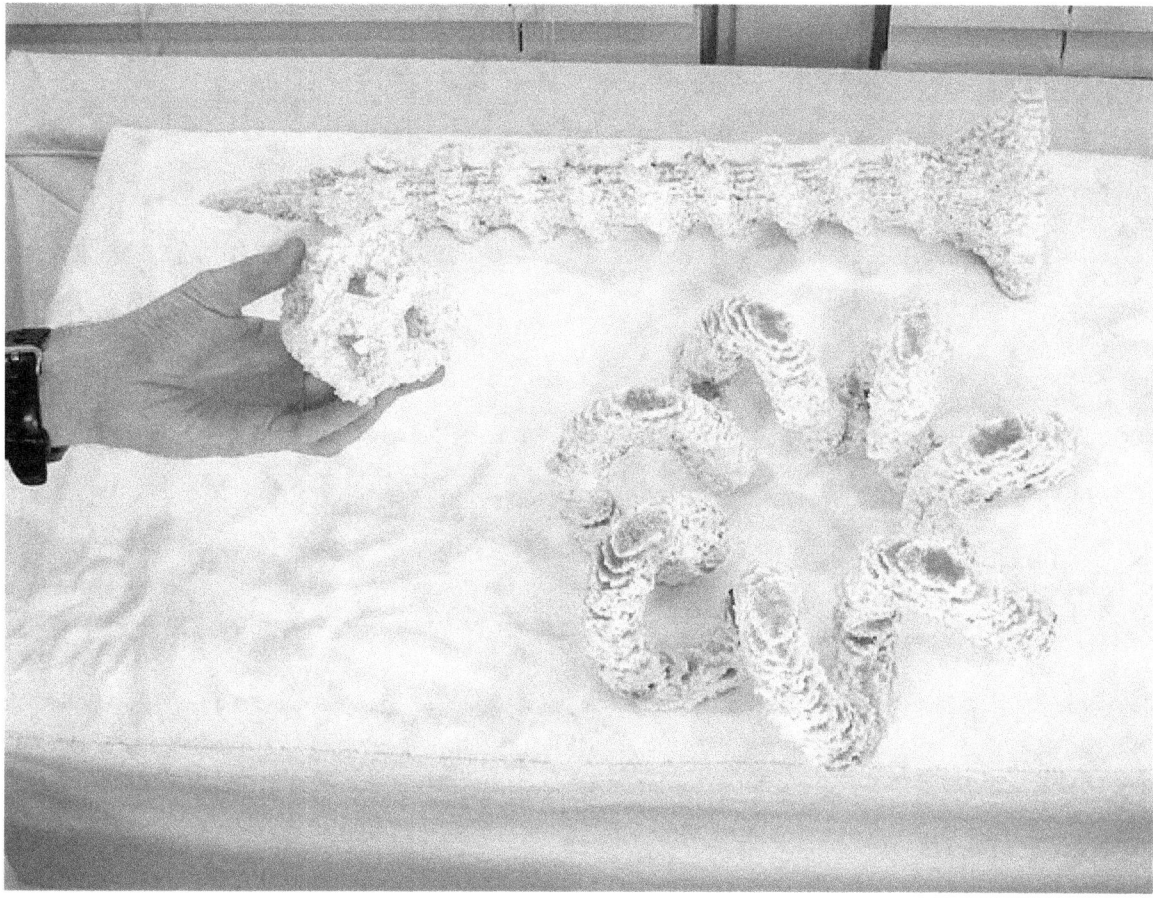

Completed toroidal coil sculpture, along with a model of a wood screw and a small dodecahedron. Each of these objects was fabricated out of pure sugar.

The prototype CandyFab 4000 unit uses granulated sugar as its print medium, giving rise to its name, but other materials with low melting temperatures and low toxicities are still under consideration.

- It uses low cost parts and construction to make it easier for others to design or build their own, with plans available as open source.

16.1 Technology

The CandyFab uses a heat source mounted on a computer-controlled X-Y positioning head to fuse the surface of a granular bed of the print media. The only thing which comes into contact with the media is heated air, which is turned on and off by the software synchronously with the motion of the positioning head. Fabrication of the shape of the part being produced progresses in layers; after each complete pass, the bed is lowered and a fresh layer of granular media is applied on top. The unfused media serves to support overhangs and thin walls in the part being produced, reducing the need for auxiliary temporary supports for the workpiece. The movable bed is of a size suitable for producing finished parts several kilograms in weight.

The resolution of features produced correspond to a smallest volume element of 2.5 x 2.5 x 2.7 mm or less.[1] Pieces produced from ordinary granular sugar have fairly good strength and feature an amber to brown surface color owing to caramelization of the sugar. Special attention has been paid to the selection of all materials coming into contact with the sugar bed or with the hot air stream to make it possible to fabricate food-grade pieces if desired.

There is an effort to encourage further work on improving the technology in the following areas:

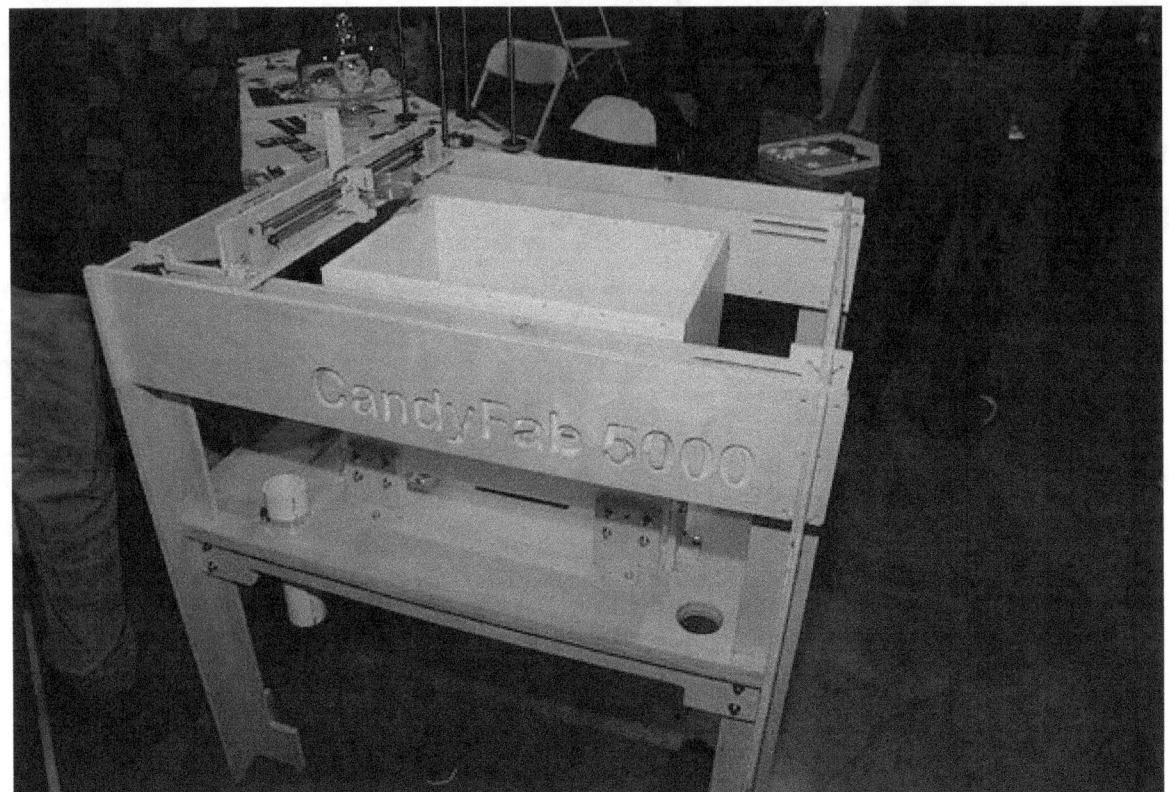

CandyFab 5000 machine at the 2008 Maker Faire event

- Hardware

- 3D modeling software

- CNC software

In addition, the inventors Windell Oskay and Lenore Edman of candyfab.org have organized teams to explore applications, gastronomy, and post processing.

16.2 Fabricated pieces

Several large pieces of sculpture have been produced using the CandyFab, including one of a mathematical object designed by sculptor Bathsheba Grossman. This and other pieces were shown by inventors Windell Oskay and Lenore Edman of candyfab.org at the Bay Area Maker Faire 2007. [2]

16.3 References

[1] Evil Mad Scientist Laboratories - Solid freeform fabrication: DIY, on the cheap, and made of pure sugar

[2] makerfaire.com:

16.4   See also

- Rapid prototyping

- Solid freeform fabrication

- Digital fabricator

16.5   External links

- The CandyFab Project

Chapter 17

Cartesian coordinate robot

A **cartesian coordinate robot** (also called **linear robot**) is an industrial robot whose three principal axis of control are linear (i.e. they move in a straight line rather than rotate) and are at right angles to each other.The three sliding joints correspond to moving the wrist up-down,in-out,back-forth. Among other advantages, this mechanical arrangement simplifies the Robot control arm solution. Cartesian coordinate robots with the horizontal member supported at both ends are sometimes called **Gantry robots**. They are often quite large.

A popular application for this type of robot is a computer numerical control machine (CNC machine) and 3D printing. The simplest application is used in milling and drawing machines where a pen or router translates across an x-y plane while a tool is raised and lowered onto a surface to create a precise design. Pick and place machines and plotters are also based on the principal of the **cartesian coordinate robot**.

17.1   See also

- Google 3D warehouse

- Sketchup

- Thingiverse

17.2   References

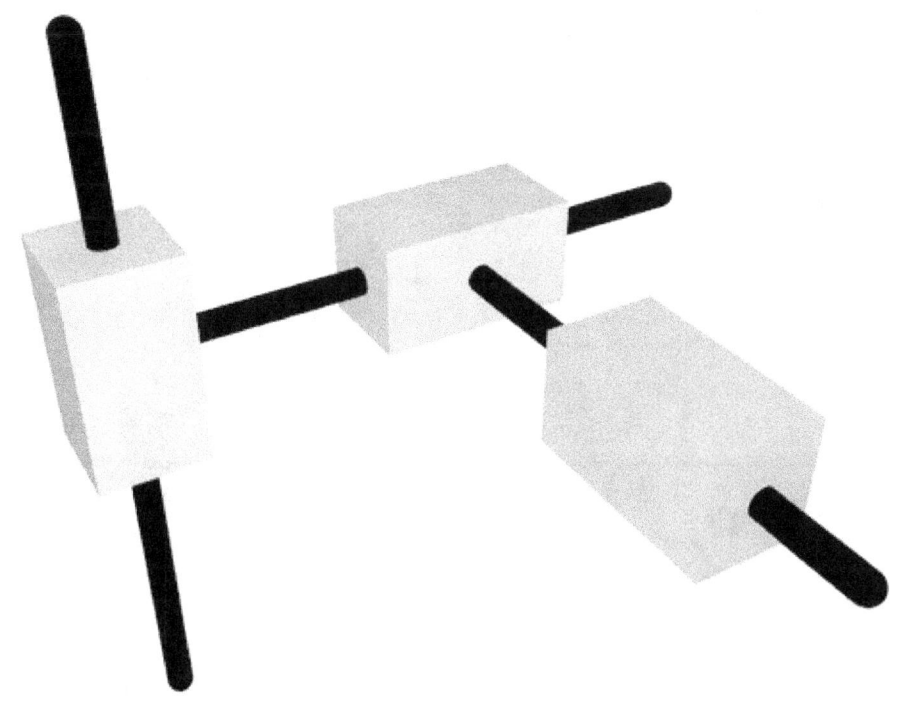

Kinematic diagram of cartesian coordinate robot

a plotter is an implemention of the cartesian coordinate robot

Chapter 18

Complexity paradox

The **complexity paradox** is a phenomenon associated with 3D printing.

The paradox arises from the observation that, in contrast with other forms of manufacturing, as the complexity of the item being manufactured increases, the cost of 3D printing the item declines.

In the case of conventional manufacturing, the more complicated an item happens to be (i.e., the more components that it is made from and the more complex the resulting assembly process happens to be) the more the item costs to make.

However, in the case of 3D printing, not only is it true that a complex 3D shape can be 'printed' just as easily as a simple one, it transpires that structural complexity actually reduces 3D printing costs

Because the cost of 3D printing an item is dependent upon little more than the amount of '3D printing ink' (as well as a small amount of electrical power) the more complex the shape happens to be, the more numerous are the spaces that there happen to be between the components (designers refer to inter-component-spaces as 'voids') the smaller the quantity of 3D printer ink (which is usually plastic, but can sometimes be metallic) that is required to create the printed object.

The 3D printing complexity paradox can be expressed in the form of an equation:

Greater complexity = more + bigger voids = less ink = lower cost

Chapter 19

Continuous Liquid Interface Production

Continuous Liquid Interface Production (**CLIP**) is a form of additive manufacturing that uses photo polymerization to create smooth-sided solid objects of a wide variety of shapes.

19.1 Process

The continuous process begins with a pool of liquid photopolymer resin. Part of the pool bottom is transparent to ultraviolet light (the "window"). An ultraviolet light beam shines through the window, illuminating the precise cross-section of the object. The light causes the resin to solidify. The object rises slowly enough to allow resin to flow under and maintain contact with the bottom of the object.[1] An oxygen-permeable membrane lies below the resin, which creates a "dead zone" (persistent liquid interface) preventing the resin from attaching to the window (photopolymerization is inhibited between the window and the polymerizer).[2]

Unlike stereolithography, the printing process is continuous. The inventors claim that it can create objects up to 100 times faster than commercial three dimensional (3D) printing methods.[1][2][3]

19.2 Applications

CLIP objects have smooth sides, unlike 2015 commercial 3D printers, whose sides are typically rough to the touch. Some resins produce objects that are rubbery and flexible, that could not be produced with earlier methods.[2]

19.3 History

19.3.1 Patents and trademarks

CLIP was, at the time the original patent was filed, an acronym for Continuous liquid interphase printing, described in two patents, titled 'Continuous liquid interphase printing' and 'Method and apparatus for three-dimensional fabrication with feed through carrier'. Both patents were filed February 10, 2014, by EiPi Systems, Inc as Applicant with the following individuals titled as 'inventors': Joseph M. DeSimone, Alexander Ermoshkin, Nikita Ermoshkin, and Edward T. Samulski.[4][5]

According to data in the California Secretary of State's office database, CARBON3D. INC is listed as of September 6, 2014.[6] A trademark was filed on September 10, 2014, for the 'CARBON3D' trademark.[7]

19.3.2   Public release

A journal article was published in *Science* detailing the groups' findings.[8] At TED 2015, Joseph M. DeSimone demonstrated a 3D-printer using CLIP technology and produced a relatively complex object in less than 10 minutes.[9] DeSimone cited a scene in the 1992 film *Terminator 2*, where the T-1000 machine reforms itself from a metallic pool, as an inspiration for the technology.[10][11]

19.4   See also

- Magnetically assisted slip casting

- Projection micro-stereolithography

19.5   References

[1] St. Fleur, Nicholas (17 March 2015). "3-D Printing Just Got 100 Times Faster". *The Atlantic*. Retrieved 19 March 2015.

[2] Castelvecchi, Davide (17 March 2015). "Chemical trick speeds up 3D printing". *Nature*. Retrieved 19 March 2015.

[3] Saxena, Shalini (19 March 2015). "New nonstop 3D printing process takes only minutes instead of hours". Ars Technica. Retrieved 19 March 2015.

[4] "Continuous liquid interphase printing". Retrieved March 20, 2015.

[5] "Method and apparatus for three-dimensional fabrication with feed through carrier". Retrieved March 20, 2015.

[6] "Business Search - Results". California Secretary of State. Retrieved March 20, 2015.

[7] "CARBON3D". United States Patent and Trademark Office. Retrieved March 20, 2015.

[8] Tumbleston, J. R.; Shirvanyants, D.; Ermoshkin, N.; Janusziewicz, R.; Johnson, A. R.; Kelly, D.; Chen, K.; Pinschmidt, R.; Rolland, J. P.; Ermoshkin, A.; Samulski, E. T.; DeSimone, J. M. (16 March 2015). "Continuous liquid interface production of 3D objects". *Science* **347** (6228): 1349–1352. doi:10.1126/science.aaa2397.

[9] "Joseph DeSimone: What if 3D printing was 100x faster?". TED. Retrieved 20 March 2015.

[10] Wakefield, Jane (17 March 2015). "TED 2015: Terminator-inspired 3D printer 'grows' objects". BBC News. Retrieved 20 March 2015.

[11] Feltman, Rachel (16 March 2015). "This mind-blowing new 3-D printing technique is inspired by 'Terminator 2'". *The Washington Post*. Retrieved 20 March 2015.

19.6   External links

- Carbon 3D website

- "See the Technology Behind a Mesmerizing, Faster 3-D Printing Process | MIT Technology Review". Retrieved 2015-07-04.

Chapter 20

D-Shape

D-Shape is a large 3-Dimensional printer that uses stereolithography, a layer by layer printing process, to bind sand with an inorganic seawater[1] and magnesium-based binder[2] in order to create stone-like objects. Invented by Enrico Dini, founder of Monolite UK Ltd, the first model of the D-Shape printer used epoxy resin, commonly used as an adhesive in the construction of skis, cars, and airplanes, as the binder. Dini patented this model in 2006.[3] After experiencing problems with the epoxy, Dini changed the binder to the current magnesium-based one and patented his printer again in September 2008.[4] In the future, Dini aims to use the printer to create full-scale buildings.

20.1   Technical Description

The current version of the D-Shape 3-D printer sits in a 6m by 6m aluminum frame. The frame consists of a square base that moves upwards along four vertical beams during the printing process via stepper motors, which are used to repeatedly move a specified length and then hold in place, on each beam. Spanning the entire horizontal 6m of the base is a printer head with 300 nozzles, each spaced 20mm apart. The printer head is connected to the base by an aluminum beam that runs perpendicular to the printer head.[5]

20.2   How it Works

Before the actual printing process can begin, a 3-D model of the object to be printed must be created on CAD, a software that allows a designer to create 3-D models on a computer. Once the model is finished, the CAD file is sent to the printer head. The printing process begins when a layer of sand from 5 to 10 mm thick, mixed with solid magnesium oxide (MgO),[6] is evenly distributed by the printer head in the area enclosed by the frame. The printer head breaks the 3-D model into 2-D slices. Then, starting with the bottom slice, the head moves across the base and deposits an inorganic binding liquid made up of a solution that includes magnesium hexahydrate, at a resolution of 25 DPI (1.0 mm).[7] The binder and sand chemically react to form a sandstone material. It takes about 24 hours for the material to completely solidify. Because the nozzles are 20mm apart there are gaps that may need to be filled up. To fill in these gaps and ensure the sand is uniformly exposed to the binder, an electric piston on the beam that holds the printer head forces the printer head to shift in the direction perpendicular to the printer's direction of motion. It takes D-Shape four forward and backward strokes to finish printing a layer. After a layer is finished, the stepper motors on the vertical beams move the base upwards. From the hollow framework just above the printer head, new sand, which is cyclically refilled, is distributed into the area of the frame to create the next layer.[8] During printing, excess sand acts as a support for the solidifying sand and can also be reused in later printings. The printing process is continuous and stops only when the desired structure is completely printed.

20.3   The End Product

After the printer finishes its work, the final structure must be extruded from the sand. Workers use shovels to take out the excess sand and reveal the final product. The magnesium oxide mixed in with the sand causes the sand to become an active participant rather than inert during the reaction with the binder. If the sand was inert, the resulting material would be more like concrete in that the sand would be only slightly bound together, but because of the MgO, the final product is a mineral-like material with a microcrystalline structure. Compared to concrete, which has low resistance to tension and as a result needs iron reinforcement, D-Shape's structures have relatively high tension resistance and require no iron reinforcement.[9] The entire building process is reported to take a quarter of the time and a third to a half of the cost[10] it would take to build the same structure with traditional means using portland cement, the material currently used in building construction.[11]

20.4   Awards and Achievements

20.4.1   NYC Waterfront Construction Competition

In the fall of 2012, D-Shape entered into the NYC Waterfront Construction Competition hosted by the New York City Economic Development Corporation (NYCEDC) in which competitors had to create an innovative solution to help strengthen New York City's deteriorating piers and coastline structures. D-Shape's idea, called, "Digital Concrete," was to take 3-D scans of each piece of pier or infrastructure, and then print a support jacket for each specific piece. D-Shape was the First Place Winner and received $50,000 for the idea, which is estimated to save New York City $2.9 billion.[12][13]

20.4.2   Radiolaria

D-Shape successfully created the tallest printed sculpture, Radiolaria, in 2009.[14] Radiolaria, a sculpture created by Italian architect Andrea Morgante and inspired by radiolarians, unicellular organisms with intricate mineral skeletons, shows off D-Shape's ability to print large freeform structures. The current version of the sculpture is only a 3 x 3 x 3m scale model of the full-size Radiolaria that is planned to be put in a roundabout in Pontedera, Italy.[15]

20.5   Future of D-Shape

Currently, Jake Wake-Walker and Marc Webb are working on a documentary, titled *The Man Who Prints Houses*, about Enrico Dini and his invention.[16] Although D-Shape has garnered attention for its printing abilities, it is still a work in progress. While it has gotten close to printing an actual house by printing a trullo, which is a small, stone hut,[17] the printer still needs to be modified in order to make Dini's dreams of printing larger and more complex buildings a reality.

20.5.1   Lunar Bases

Because of D-Shape's capabilities, the European Space Administration (ESA) has taken interest in using the printer to build moon bases.[18] The ESA is interested in using D-Shape to build moon bases out of lunar regolith, otherwise known as moon dust, because the 3-D printer can build the base onsite without human intervention. This is advantageous because only the machine would have to be taken to the moon, thus reducing the cost of bringing building materials to the lunar surface to create the bases. D-Shape has been successful in printing components for the lunar bases with a simulated moon dust, and has also been subject to tests that aim to see how the printer will work in the environment on the moon.[19]

20.6   References

[1]   "Discovery Channel Covers DShape 3D Printing". Youtube, DShape3DPrinting. Retrieved 21 October 2013.

[2] Cesaretti, Giovanni; Enrico Dini; Xavier de Kestelier; Valentina Colla; Laurent Pambaguian (January 2014). "Building components for an outpost on the Lunar soil by means of a novel 3D printing technology". *Acta Astronautica* **93**: 430–450. doi:10.1016/j.actaastro.2013.07.034. Retrieved 4 November 2013.

[3] Dini, Enrico. "Method and device for building automatically conglomerate structures CA 2602071 A1". US Patents. Retrieved 11 November 2013.

[4] Dini, Enrico. "Method for automatically producing a conglomerate structure and apparatus therefor US 8337736 B2". US Patents. Retrieved 11 November 2013.

[5] Cesaretti, Giovanni; Enrico Dini; Xavier de Kestelier; Valentina Colla; Laurent Pambaguian (January 2014). "Building components for an outpost on the Lunar soil by means of a novel 3D printing technology". *Acta Astronautica* **93**: 430–450. doi:10.1016/j.actaastro.2013.07.034. Retrieved 4 November 2013.

[6] Cesaretti, Giovanni; Enrico Dini; Xavier de Kestelier; Valentina Colla; Laurent Pambaguian (January 2014). "Building components for an outpost on the Lunar soil by means of a novel 3D printing technology". *Acta Astronautica* **93**: 430–450. doi:10.1016/j.actaastro.2013.07.034. Retrieved 4 November 2013.

[7] Edwards, Lin (19 April 2010). "3D printer could build moon bases". *Phys.org*. Retrieved 21 October 2013.

[8] Cesaretti, Giovanni; Enrico Dini; Xavier de Kestelier; Valentina Colla; Laurent Pambaguian (January 2014). "Building components for an outpost on the Lunar soil by means of a novel 3D printing technology". *Acta Astronautica* **93**: 430–450. doi:10.1016/j.actaastro.2013.07.034. Retrieved 4 November 2013.

[9] Dini, Enrico. "Method for automatically producing a conglomerate structure and apparatus therefor US 8337736 B2". US Patents. Retrieved 11 November 2013.

[10] Parsons, Sarah (17 March 2010). "3-D Printer Creates Entire Buildings From Solid Rock". *Habitat*. Retrieved 22 October 2013.

[11] Belezina, Jan (24 February 2012). "D-Shape 3D printer can print full-sized houses". *Gizmag*. Retrieved 21 October 2013.

[12] "D-Shape Promises To Modernize New York's Shoreline Using 3D-Printing Technology". *The Huffington Post*. 3 June 2013. Retrieved 21 October 2013.

[13] "D-Shape wins top prize in NYC Waterfront Construction Competition". *3ders.org*. 12 April 2013. Retrieved 20 October 2013.

[14] Quirk, Vanessa. "How 3D Printing Will Change Our World". Arch Daily. Retrieved 20 October 2013.

[15] Edwards, Lin (19 April 2010). "3D printer could build moon bases". *Phys.org*. Retrieved 21 October 2013.

[16] Blagdon, Jeff (21 February 2012). "British company uses 3D printing to make stone buildings out of sand". *The Verge*. Retrieved 21 October 2013.

[17] Quirk, Vanessa. "How 3D Printing Will Change Our World". Arch Daily. Retrieved 20 October 2013.

[18] Edwards, Lin (19 April 2010). "3D printer could build moon bases". *Phys.org*. Retrieved 21 October 2013.

[19] Cesaretti, Giovanni; Enrico Dini; Xavier de Kestelier; Valentina Colla; Laurent Pambaguian (January 2014). "Building components for an outpost on the Lunar soil by means of a novel 3D printing technology". *Acta Astronautica* **93**: 430–450. doi:10.1016/j.actaastro.2013.07.034. Retrieved 4 November 2013.

20.7 External links

- D-Shape's official Website

- Discovery Channel Covering D-Shape http://www.youtube.com/watch?v=RYaRUVTwIVc

Chapter 21

Defense Distributed

Defense Distributed is an online, open-source[1] organization that designs firearms, or "wiki weapons",[5][6][7] that may be downloaded from the Internet and "printed" with a 3D printer.[5] Among the organization's goals is to develop and freely publish firearms-related design schematics that can be downloaded and reproduced by anyone with a 3D printer.[8][9]

After raising over US$20,000 via a crowd-funding appeal,[5][9] suffering the confiscation of its first 3D printer,[10] and partnering with private manufacturing firms,[11] the organization began live fire testing of printable firearm components in December 2012.[12][13]

Defense Distributed has to date produced a durable printed receiver for the AR-15,[14][15][16] the first printed standard capacity AR-15 magazine,[17][18][19] and the first printed magazine for the AK-47.[20][21] These 3D printable files were available for download at the organization's former publishing site, DEFCAD,[22] but are now largely hosted on file sharing websites.[23][24]

On May 5, 2013, Defense Distributed made public the 3D printable files (STL files) for the world's first fully 3D printable gun, the Liberator .380 single shot pistol.[25][26][27] Days later, the United States Department of State demanded the files be removed from the Internet, citing a violation of the International Traffic in Arms Regulations.[28][29]

On May 6, 2015, Defense Distributed, joined by the Second Amendment Foundation, brought suit against the Department of State in the western district of Texas.[30][31]

21.1 History

21.1.1 Founding

The defensedistributed.com domain name was registered on June 4, 2012.[3] The website was unveiled in conjunction with an Indiegogo campaign of the same name in July 2012, where the organization asked to receive US$20,000.[5][32] Indiegogo suspended the crowd-funding campaign for a terms of service violation after three weeks, refunding the money raised without offering public comment.[32][33] Defense Distributed continued the appeal on its own website, however, accepting contributions through PayPal and the cryptocurrency Bitcoin, and met its fundraising goal in September 2012.[34]

The organization has been predominantly represented in public since July 2012 by Cody Wilson, who is described as a founder and spokesperson.[7][35]

Defense Distributed lists its members as a mix of students, IT professionals, engineers, and programmers from the United States and Germany.[1]

21.1.2 Purpose

According to the Defense Distributed website, the nonprofit is organized and operated for charitable and literary purposes, specifically "to defend the civil liberty of popular access to arms as guaranteed by the United States Constitution and affirmed by the Supreme Court, through facilitating global access to, and the collaborative production of, information and knowledge related to the 3D printing of arms; and to publish and distribute... such information and knowledge in promotion of the public interest."[1][11] The website's "Manifesto" link directs users to an online version of John Milton's essay *Areopagitica*.[36]

The organization's motivations have been described as "less about [a] gun... than about democratizing manufacturing technology,"[37] In an interview with *Slashdot*, Cody Wilson described the Wiki Weapon project as a chance to "experiment with Enlightenment ideas… to literally materialize freedom."[38]

At Bitcoin 2012 in London, Wilson explained the organization as interested in inspiring libertarian forms of social organization and technologically driven inversions of authority.[39]

21.1.3 DEFCAD

Main article: DEFCAD

In December 2012, as a response to Makerbot Industries' decision[40][41][42] to remove firearms-related 3D printable files at the popular repository Thingiverse, Defense Distributed launched a companion site at defcad.org to publicly host the removed 3D printable files and its own.[43][44][45] Public and community submissions to DEFCAD rose quickly,[22][45][46] and in March 2013, at the SXSW Interactive festival, Wilson announced a repurposed and expanded DEFCAD as a separate entity that would serve as a 3D search engine and development hub, while maintaining the spirit of access endemic to Defense Distributed.[47][48][49] The new DEFCAD was deemed "The Pirate Bay of 3D Printing"[50] and "the anti-Makerbot"[49] even before its launch, provided an index of over 100,000 files.[51]

21.1.4 Ghost Gunner

In October 2014, Defense Distributed began selling to the public a miniature CNC mill for completing receivers for the AR-15 semi-automatic rifle.[52] For a review of the machine in Wired, Andy Greenberg manufactured a series of lowers and called the machine "absurdly easy to use."[53]

21.2 Administration

21.2.1 Legal History

Defense Distributed is a pending 501(c)(3) federal tax exempt organization, and not a weapons manufacturer.[7][11][13] The organization operates to publish intellectual property and information developed by licensed firearms manufacturers and the public.[11]

Cody Wilson has a Type 7 Federal Firearms License (FFL), however.[54][55]

21.3 Legal Challenges

21.3.1 Stratasys confiscation

Learning of Defense Distributed's plans in 2012, manufacturer Stratasys, Ltd threatened legal action and demanded the return of the 3D printer it had leased to Wilson.[10] On September 26, before the printer was assembled for use, Wilson received an email from Stratasys suggesting that he might use the printer "for illegal purposes".[10] Stratasys immediately canceled its lease with Wilson and sent a team to confiscate the printer the next day.[10][13] Wilson was subsequently questioned by the ATF when visiting an ATF field office in Austin, Texas to inquire about legalities and regulations relating to the Wiki Weapons project.[10]

21.3.2 The Undetectable Firearms Act

Defense Distributed's efforts have prompted renewed discussion and examination of the Undetectable Firearms Act.[7][55][56][57] The Liberator pistol was cited in White House and Congressional calls to renew the Act in 2013.[58][59]

21.3.3 International Traffic in Arms Regulations

On May 9, 2013, The United States Department of State Directorate of Defense Trade Controls (DDTC) directed Defense Distributed to remove the download links to its publicly accessible CAD files.[60] The State Department's letter, likely prompted by the Liberator Pistol, referenced § 127.1 of the International Traffic in Arms Regulations (ITAR), interpreting the regulations to impose a prior approval requirement on publication of Defense Distributed's files into the public domain,[61] a legal position noted at the time to suffer from First and Second Amendment infirmities.[61][62]

On May 6, 2015, Defense Distributed filed a Constitutional challenge against the State Department in the Western District of Texas, suing government agents within the DDTC and accusing the government of knowingly violating the company's First and Second and Fifth Amendment liberties.[30][31] Defense Distributed was joined in its suit by the Second Amendment Foundation.[31]

As of September 2015, the case is on an expedited interlocutory appeal before the United States Court of Appeals for the Fifth Circuit.[63]

Peer-to-peer torrent sites and other repositories continue to host Defense Distributed and other firearms CAD files.[64][28][29]

21.4 Reception

Defense Distributed has been obliquely endorsed by the Gun Owners of America (GOA).[65] However the National Rifle Association (NRA) has offered - to date - no public comment on the organization or its activities.

Open source software advocate Eric S. Raymond has endorsed the organization and its efforts, calling Defense Distributed "friends of freedom" and writing "I approve of any development that makes it more difficult for governments and criminals to monopolize the use of force. As 3D printers become less expensive and more ubiquitous, this could be a major step in the right direction."[66][67]

Aaron Timms of Blouin News has written Defense Distributed has performed "the greatest piece of political performance art of [the 21st] century.",[68]

For its activities, Defense Distributed has been accused of endangering public safety and attempting to frustrate and alter the US system of government.[69][70]

21.5 See also

- 3D printed firearms
- Gun control
- Gun politics in the United States
- Improvised firearm
- List of notable 3D printed weapons and parts

United States Department of State

Bureau of Political-Military Affairs
Office of Defense Trade Controls Compliance
Washington, D.C. 20522-0112

MAY 0 8 2013

In reply refer to

████████████████

Mr. Cody Wilson
Defense Distributed
████████████████

Dear Mr. Wilson:

The Department of State, Bureau of Political Military Affairs, Office of Defense Trade Controls Compliance, Enforcement Division (DTCC/END) is responsible for compliance with and civil enforcement of the Arms Export Control Act (22 U.S.C. 2778) (AECA) and the AECA's implementing regulations, the International Traffic in Arms Regulations (22 C.F.R. Parts 120-130) (ITAR). The AECA and the ITAR impose certain requirements and restrictions on the transfer of, and access to, controlled defense articles and related technical data designated by the United States Munitions List (USML) (22 C.F.R. Part 121).

DTCC/END is conducting a review of technical data made publicly available by Defense Distributed through its 3D printing website, DEFCAD.org, the majority of which appear to be related to items in Category I of the USML. Defense Distributed may have released ITAR-controlled technical data without the required prior authorization from the Directorate of Defense Trade Controls (DDTC), a violation of the ITAR.

Technical data regulated under the ITAR refers to information required for the design, development, production, manufacture, assembly, operation, repair, testing, maintenance or modification of defense articles, including information in the form of blueprints, drawings, photographs, plans, instructions or documentation. For a complete definition of technical data, see § 120.10 of the ITAR. Pursuant to § 127.1 of the ITAR,

Letter from the United States Department of State to Defense Distributed (May 8, 2013).

21.6   References

[1] "About Us". Defense Distributed. Retrieved December 15, 2012.

[2] "Defense Distributed". Defense Distributed. Retrieved October 11, 2012.

[3] "Whois Search Results: defensedistributed.com". Retrieved September 21, 2012.

[4] "Defdist.org Site Info". Alexa Internet. Retrieved July 2, 2015.

[5] Greenberg, Andy (August 23, 2012). "'Wiki Weapon Project' Aims To Create A Gun Anyone Can 3D-Print At Home". *Forbes*. Retrieved August 27, 2012.

[6] Bilton, Nick (October 7, 2012). "Disruptions: With a 3-D Printer, Building a Gun With the Push of a Button". *The New York Times*. Retrieved December 15, 2012.

[7] Doherty, Brian (December 12, 2012). "Disruptions: With a 3-D Printer, Building a Gun With the Push of a Button". *Reason.com*. Retrieved December 15, 2012.

[8] Hobbyist builds working assault rifle using 3D printer

[9] Poeter, Damon (August 24, 2012). "Could a 'Printable Gun' Change the World?". *PC Magazine*. Retrieved August 27, 2012.

[10] Beckhusen, Robert (October 1, 2012). "3-D Printer Company Seizes Machine From Desktop Gunsmith". Wired News. Retrieved October 4, 2012.

[11] Hotz, Alexander (November 25, 2012). "3D 'Wiki Weapon' guns could go into testing by end of year, maker claims". *The Guardian*. Retrieved December 15, 2012.

[12] Beckhusen, Robert (December 3, 2012). "3-D Printed Gun Only Lasts 6 Shots". *Wired*. Retrieved December 15, 2012.

[13] Greenberg, Andy (December 3, 2012). "Here's What It Looks Like To Fire A (Partly) 3D-Printed Gun (Video)". *Forbes Online*. Retrieved December 15, 2012.

[14] Beckhusen, Robert (February 28, 2013). "Watch the New and Improved Printable Gun Spew Hundreds of Bullets". *Wired*. Retrieved April 12, 2013.

[15] Farivar, Cyrus (March 1, 2013). ""Download this gun": 3D-printed semi-automatic fires over 600 rounds". *Ars Technica*. Retrieved April 12, 2013.

[16] Biggs, John (March 1, 2013). "Defense Distributed Prints An AR-15 Receiver That Has Fired More Than 600 Rounds". *TechCrunch*. Retrieved April 12, 2013.

[17] Greenberg, Andy (January 14, 2013). "Gunsmiths 3D-Print High Capacity Ammo Clips To Thwart Proposed Gun Laws". *Forbes Online*. Retrieved April 12, 2013.

[18] Franzen, Carl (February 7, 2013). "Defense Distributed Unveils New 3D Printed Gun Magazine 'Cuomo' (VIDEO)". *Talking Points Memo*. Retrieved April 12, 2013.

[19] Beckhusen, Robert (February 8, 2013). "New 3-D Printed Rifle Magazine Lets You Fire Hundreds of Rounds". *Wired Danger Room*. Retrieved April 12, 2013.

[20] Ingersoll, Geoffrey (March 8, 2013). "3D Printing Company Names AK-47 Magazine After Gun Control Congresswoman". *Business Insider*. Retrieved April 12, 2013.

[21] Branson, Michael (April 8, 2013). "Defense Distributed Releases Printable AK Magazine". *The Firearm Blog*. Retrieved April 12, 2013.

[22] Bilton, Ricardo (February 19, 2013). "3D-printing gun site DEFCAD now attracting 3K visitors an hour, 250K downloads since launch". *VentureBeat*. Retrieved April 12, 2013.

[23] "Defiant Pirate Bay to continue hosting banned 3D printer gun designs". *RT.com*. 10 May 2013. Retrieved 4 August 2013.

[24] Ernesto. "Pirate Bay Takes Over Distribution of Censored 3D Printable Gun". *TorrentFreak*. Retrieved 13 May 2013.

[25] Greenberg, Andy (May 5, 2013). "Meet The 'Liberator': Test-Firing The World's First Fully 3D-Printed Gun". *Forbes*. Retrieved May 7, 2013.

[26] Morelle, Rebecca (May 6, 2013). "Working gun made with 3D printer". *BBC News*. Retrieved 28 July 2013.

[27] Hutchinson, Lee. "The first entirely 3D-printed handgun is here". *Ars Technica*. Retrieved 13 May 2013.

[28] "3D-printed gun blueprints pulled from Internet, at request of State Department". *CBS News*. May 10, 2013. Retrieved May 10, 2013.

[29] Nozowitz, Dan. "U.S. State Department Tells Defense Distributed To Take Down 3-D Printed Gun Plans". Popular Science. Retrieved May 10, 2013.

[30] Feuer, Alan (May 6, 2015). "Cody Wilson, Who Posted Gun Instructions Online, Sues State Department". *The New York Times*. Retrieved August 29, 2015.

[31] Greenberg, Andy (May 6, 2015). "3-D Printed Gun Lawsuit Starts the War Between Arms Control and Free Speech". *Wired*. Retrieved August 29, 2015.

[32] Roy, Jessica (August 23, 2012). "WikiWeapon Campaign to 3D-Print Your Own Gun Suspended by Indiegogo". *Betabeat*. Retrieved December 15, 2012.

[33] Martinez, Fidel (August 27, 2012). "Indiegogo shuts down campaign to develop world's first printable gun". *The Daily Dot*. Retrieved December 15, 2012.

[34] Greenberg, Andy (September 20, 2012). "3D-Printable Gun Project Hits Its Fundraising Goal Despite Being Booted Off Indiegogo". *Forbes Online*. Retrieved December 15, 2012.

[35] Brown, Rich (September 7, 2012). "You don't bring a 3D printer to a gun fight - yet - Yahoo! News". News.yahoo.com. Retrieved October 6, 2012.

[36] "Want a Free Download of a Semi-Automatic Rifle? Print One!". thelibertarianrepublic.com. March 3, 2013. Retrieved April 12, 2013.

[37] Brown, Rich (September 6, 2012). "You don't bring a 3D printer to a gun fight -- yet". *CNET*. Retrieved September 21, 2012.

[38] "Should We Print Guns? Cody R. Wilson Says "Yes" (Video) -Slashdot". Hardware.slashdot.org. Retrieved October 6, 2012.

[39] "Bitcoin2012 London". Bitcoin2012.com. Retrieved October 6, 2012.

[40] Maly, Tim (December 19, 2012). "Thingiverse Removes (Most) Printable Gun Parts". Wired. Retrieved January 14, 2013.

[41] "MakerBot pulls 3D gun-parts blueprints after Sandy Hook". BBC News. December 20, 2012. Retrieved January 14, 2013.

[42] Pepitone, Julianne (December 20, 2012). "3-D printer MakerBot cracks down on blueprints for gun parts". CNN Money. Retrieved January 14, 2013.

[43] Limer, Eric (December 21, 2012). "There's a New Site Just for 3D-Printed Gun Designs". Gizmodo. Retrieved January 14, 2013.

[44] Bilton, Ricardo (December 21, 2012). "Fighting 'censorship,' 3D-printed gun designs find a new home". VentureBeat. Retrieved January 14, 2013.

[45] Robertson, Adi (December 21, 2012). "3D printed gun enthusiasts build site for firearm 3D printable files after MakerBot crackdown". The Verge. Retrieved January 14, 2013.

[46] Klimas, Liz (January 9, 2012). "Website to The Blaze: People Rushing to Download Online Blueprints for 3D Printed Guns". The Blaze. Retrieved January 14, 2013.

[47] Greenberg, Andy (March 11, 2013). "3D-Printable Gun Project Announces Plans For A For-Profit Search Engine Startup". *Forbes Online*. Retrieved April 12, 2013.

[48] Farivar, Cyrus (March 11, 2013). "3D printing gunmaker forms company to flout copyright law, à la the Pirate Bay". *Ars Technica*. Retrieved April 12, 2013.

[49] Bilton, Ricardo (March 11, 2013). "Expanding beyond 3D printed guns, DEFCAD is officially the anti-MakerBot". *VentureBeat*. Retrieved April 12, 2013.

[50] "'Pirate Bay' for 3D printing launched". *BBC News*. March 12, 2013. Retrieved April 12, 2013.

[51] "DEFCAD.com". DEFCAD. Retrieved August 6, 2013.

[52] Greenberg, Andy (October 1, 2015). "The $1,200 Machine That Lets Anyone Make a Metal Gun at Home". Wired. Retrieved September 2, 2015.

[53] Greenberg, Andy (June 3, 2015). "I Made an Untraceable AR-15 'Ghost Gun' in My Office—And It Was Easy". Wired. Retrieved September 2, 2015.

[54] "US grants first license to sell 3D-printed guns". *Daily Mail*. March 18, 2013. Retrieved March 21, 2013.

[55] LeJacq, Yannick (December 10, 2012). "Defense Distributed's 'Wiki Weapon': U.S. Congressman Steve Israel Offers First Legislative Challenge". Retrieved December 15, 2012.

[56] Hsu, Jeremy (December 10, 2012). "3D-Printable Guns Face Federal Ban". *Mashable*. Retrieved December 15, 2012.

[57] Brown, Rich (December 10, 2012). "The Undetectable Firearms Act and 3D-printed guns (FAQ)". *CNET*. Retrieved December 15, 2012.

[58] Pérez, Evan (November 15, 2013). "ATF tests show 3-D guns lethal as metal detection law expires". *CNN.com*. Retrieved May 16, 2014.

[59] Schmidt, Michael (November 28, 2013). "Law Limiting Plastic Guns Set to Expire". *The New York Times*. Retrieved May 16, 2014.

[60] Preston, Jennifer (May 10, 2013). "Printable-Gun Instructions Spread Online After State Dept. Orders Their Removal". *The New York Times*. Retrieved May 16, 2014.

[61] Morris, Kevin (September 27, 2013). "The Liberator: Cody Wilson's armed for a free speech battle". *ValleyWag*. Retrieved May 16, 2014.

[62] Goldstein, Matthew (June 15, 2013). "Department of State Confirms Prior Approval Requirement for Electronic Exports to Public Domain in Case of 3D-Printable Gun". *Thomson Reuters Practical Trade & Customs Strategies* (Thomson Reuters) **2** (11): 3–6. Retrieved May 17, 2014.

[63] Doherty, Brian (August 7, 2015). "Defense Distributed Injunction Request Denied in Suppression of Gun-Related Internet Speech Case". Wired. Retrieved September 2, 2015.

[64] Greenberg, Andy (May 9, 2013). "State Department Demands Takedown Of 3D-Printable Gun Files For Possible Export Control Violations". *Forbes*. Retrieved May 10, 2013.

[65] Rosenwald, Michael (February 18, 2013). "Weapons made with 3-D printers could test gun-control efforts". *The Washington Post*.

[66] Raymond, Eric (August 23, 2012). "Defense Distributed". Armed and Dangerous. Retrieved January 14, 2013.

[67] Kopstein, Joshua (April 12, 2013). "Guns want to be free: what happens when 3D printing and crypto-anarchy collide?". *The Verge*.

[68] Timms, Aaron (March 29, 2013). "The future of 3D printing might be scarier than you thought". *Blouin News*.

[69] "The 15 Most Dangerous People in the World". Wired Danger Room. December 19, 2012. Retrieved January 14, 2013.

[70] Morozov, Evgeny (March 16, 2013). "Open and Closed". *The New York Times*. Retrieved April 12, 2013.

21.7   External links

- Defense Distributed's official website

- DEFCAD

- The Wiki Weapon development blog

Chapter 22

DFM analysis for stereolithography

Main article: Stereolithography

Stereolithography is an Additive manufacturing process. In this process, parts are built from a photo-curable liquid resin that cures when exposed to a laser beam (basically-undergoing the photo-polymerization process), which scans across the surface of the resin. Resins containing acrylate, epoxy, and urethane are typically used. Complex parts and assemblies can be directly made in one go unlike conventional stages of manufacturing such as casting, forming, joining and machining. Realization of such a seamless process requires the designer to take in considerations of manufacturability of the part (or assembly) by the process. One such tool is Design for manufacturability (DFM) analysis. In any product design process, DFM considerations are important to reduce iterations, time and material wastage. Described here are DFM considerations while designing a part (or assembly) to be manufactured by Stereolithography process.

22.1 Challenges in Stereolithography

22.1.1 Material

Excessive setup specific material cost and lack of support for 3rd party resins is a major challenge with SLA process:.[1] The choice of material (a design process) is restricted by the supported resin. Hence, the mechanical properties are also fixed. When scaling up dimensions selectively to deal with expected stresses, post curing is done by further treatment with UV light and heat.[2] Although advantageous to mechanical properties, the additional polymerization and cross linkage can result in shrinkage, warping and residual thermal stresses.[3] Hence, the part shall be designed in its 'green' stage i.e. pre-treatment stage.

22.1.2 Setup and process

SLA process is an additive manufacturing process. Hence, design considerations such as orientation, process latitude, support structures etc. have to be considered.[4] Orientation affects the support structures, manufacturing time, part quality and part cost.[5] Complex structures may fail to manufacture properly due to orientation which is not feasible resulting in undesirable stresses. This is when the DFM guidelines can be applied. Design feasibility for stereolithography can be validated by analytical [6] as well as on the basis of simulation and/or guidelines [7]

22.2 Rule-based DFM Considerations

Rule-based considerations in DFM refer to certain criteria that the part has to meet in order to avoid failures during manufacturing. Given the layer-by-layer manufacturing technique the process follows, there isn't any constraint on the

A desktop setup for Rapid prototyping by SLA process.

overall complexity that the part may have. But some rules have been developed through experience by the printer developer/academia which must be followed to ensure that the individual features that make up the part are within certain 'limits of feasibility'.

22.2.1 Printer constraints

Constraints/limitations in SLA manufacturing comes from the printer's accuracy, layer thickness, speed of curing, speed of printing etc. Various printer constraints are to be considered during design such as:[8]

- **Minimum Wall Thickness (Supported and Unsupported)**: Wall thickness in geometries is limited by resin resolution. Supported walls have ends connected to other walls. Below a thickness limit, such walls wall may warp during peeling. Unsupported walls are even more liable to detachment hence higher limit is for such case.

- **Overhang (Maximum Unsupported Length and Minimum Unsupported Angle)**: Overhangs are geometric features that are not supported inherently in the part. These must be supported by support structures. There is a maximum limit when structures are not provided. This is to reduce bending under self-wight. Too shallow angles result in a longer unsupported (projected) length. Hence, a minimum limit on that.

- **Maximum Bridge Span**: To avoid sagging of beam-like structures that are supported only at the ends, the maximum span length of such structures shall be limited. Whenever this is not possible, width should be increased for compensation.

- **Minimum Vertical pillar diameter**: This is to ensure the slenderness is above a limit at which the feature becomes wavy.

- **Minimum dimensions of grooves and embossed detail**: Grooves are imprinted and emboss are shallow raised features on the part surface. Features printed with dimensions smaller than the limits are unrecognizable.

- **Minimum Clearance between geometries**: This is to ensure the parts don't fuse.

- **Minimum hole diameter and radius of curvatures**: Small curvatures that aren't realizable by print dimensions may close up or smooth out/fuse.

- **Minimum internal volumes nominal diameters**: Volumes that are too small may fill up.

22.2.2 Support Structures

A point needs support if:[9]

- It is end point of support less edges

- If length of the overhang is more than a critical value

- It is at the geometric center of support less plane

While printing, support structures act as a part of design hence, their limitations and advantages are kept in mind while designing. Major considerations include:

- **Support shallow angle geometry**: Shallow angles may result in improper resin (structural strength issues) curing unless supports are provided uniformly. Generally, beyond a certain angle (usually around 45 degrees), the surface doesn't require support.

- **Overhang base**: Increase section thickness at base to avoid tearing. Avoid sharp transitions at overhang base.

- **Air pocket releaf**: Without supports, printing parts with a flat surface and holes in the geometry may create air bubbles. As the part prints, these air pockets can cause voids in the model. The support structures, in this case, create pathways through which the air bubbles could escape.[10]

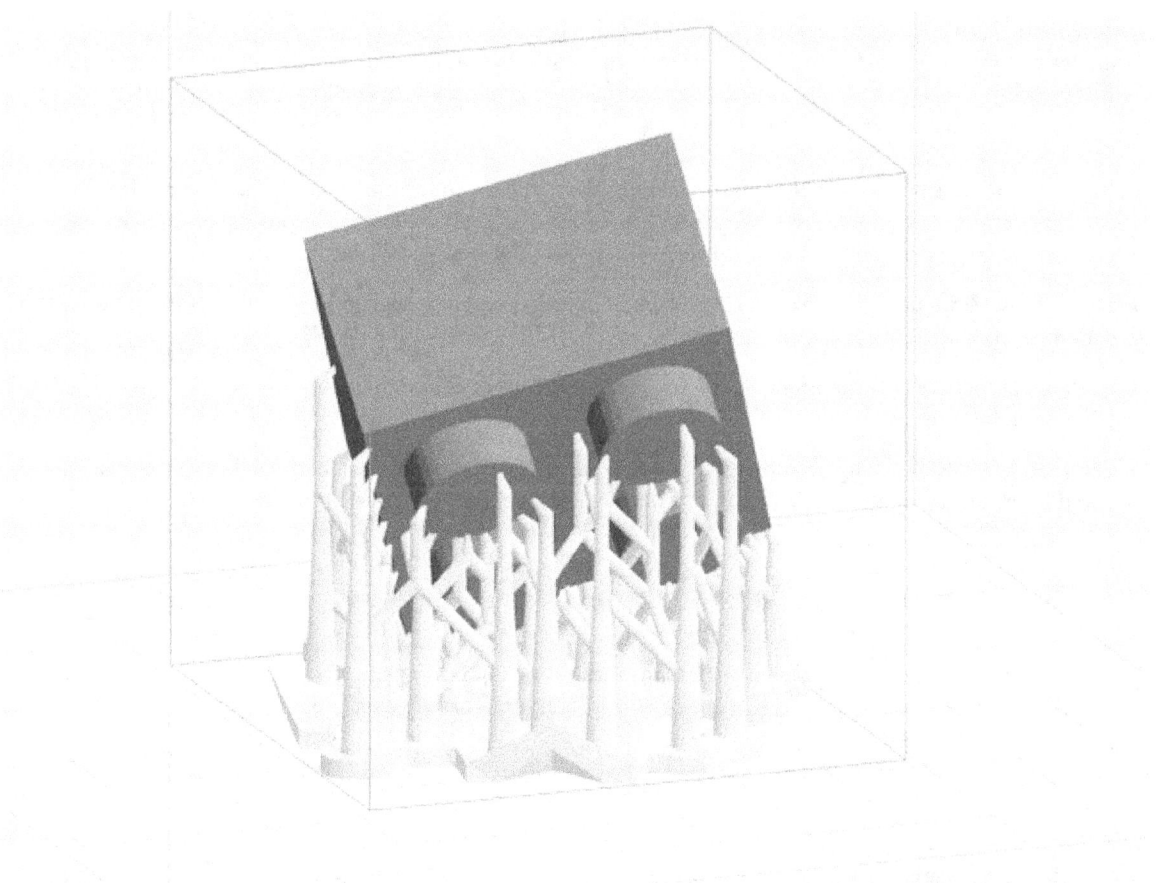

Graphic showing support structures for a lego block

- **Structure compatibility**: Consider Supports compatibility for internal volume surface.

- **Feature Orientation**: Orient to ensure overhangs are well supported.

22.2.3   Part deposition orientation

Significance of support structures and orientation in SLA process. The object, in first case, has strength issues and takes more time to manufacture than in the second case.

Part orientation is a very crucial decision in DFM analysis for SLA process. The build time, surface quality, volume/number of support structures etc. depend on this. In many cases, it is also possible to address the manufacturability issues just by reorienting the part. For example, an overhanging geometry with shallow angle may be oriented to ensure steep angles. Hence, major considerations include:

- **Surface finish improvement**: Orient the part in such a way that a feature on critical surface is eliminated. Algorithmic point of view, a free-form surface is decomposed to combination of various plane surfaces and weight is calculated/assigned to each. Total of weights is minimized for best overall surface finish.[9]

- **Build Time reduction**: Rough estimation of build time is done using slicing. The build time is proportional to the sum of surface areas of each slice. (Can be approximated as height of the part)

- **Support structure optimization**: Supported area varies as per orientation. In some orientations, it is possible to reduce support area.

- **Easy peel-off**: Reorienting such that the projected area of layers varies gradually makes it easier to peel off the cured layer during printing. Orientation also helps in removal of the support structures at later stages.

22.3 Plan-based DFM Considerations

Plan-based considerations in DFM refer to criteria that arise due to process plan. These are to be met in order to avoid failures during manufacturing of a part that may be satisfy the rule-based criteria but may have some manufacturing difficulties due to sequence in which features are produced.

22.3.1 Geometric Tailoring

"The modification of some non-critical geometric features of a part to lower fabrication cost and time, and to produce functional prototypes that mimic the behavior of the production parts.[11]"

Geometric Tailoring bridges the mismatch of material properties and process differences described above. Both functionality and manufacturability issues are addressed. Functionality issues are addressed through 'tailoring' of dimensions of the part to compensate the stress and deflection behavior anomalies.[11] Maufacturability issues are tackled through identification of difficult to manufacture geometric attributes (an approach used in most DFM handbooks) or through simulations of manufacturing processes. For RP-produced parts (as in SLA), the problem formulations are called material-process geometric tailoring (MPGT)/RP. First, the designer specifies information such as: Parametric CAD model of the part; constraints and goals on functional, geometry, cost and time characteristics; analysis models for these constraints and goals; target values of goals; and preferences for the goals. DFM problem is then formulated as the designer fills in the MPGT template with this information and sends to the manufacturer, who fills in the remaining 'manufacturing relevant' information. With the completed formulation, the manufacturer is now able to solve the DFM problem, performing GT of the part design. Hence, the MPGT serves as the digital interface between the designer and the manufacturer. Various Process Planning (PP) strategies have been developed for geometric tailoring in SLA process.[12][13]

22.3.2 DFM Frameworks

The constraints imposed by the manufacturing process are mapped onto the design. This helps in identification of DFM problems while exploring process plans by acting as a retrieval method. Various DFM frameworks are developed in literature. These frameworks help in various decision making steps such as:

- Product-process fit: Ensuring consideration of manufacturing issues during the design stage gives insight on whether SLA process is the right choice. Rapid prototyping can be done in various ways. The usual concern are process cost and availability. Through this DFM Framework, the designer can make necessary design changes to ease the component manufacturability in SLA Process.[14] This framework hence ensures that the product is suitable for the manufacturing plan.

- Feature recognition: This is done through integrated process planning tasks in commercial CAD/CAM software. This may include simulations of the manufacturing process to get an idea of the possible difficulties in a virtual manufacturing environment. Such integrated tools are in developmental stage.

- Functionality considerations: In some cases, assemblies are directly printed instead of printing parts separately and assembling. In such cases, phenomenon such as flow of the resin may affect the functionality drastically which may not be addressed through just rule based analysis. In fact, the rule based analysis is only to ensure the bounds of design but the dimensions of the final part must be checked for manufacturability through Plan-based consideration. Considerable research has been going on in this since the past decade.[15][16] DFM frameworks are being developed and put into packages.[17]

22.4   See also

- Stereolithography

- Design for manufacturability

- Rapid prototyping

22.5   References

[1] 3D printing issues and challenges: Material costs

[2] Bártolo, Paulo. *Stereolithography: Materials, Processes and Applications*. Springer, 2011, p. 130

[3] D Karalekas, A Aggelopoulos, "Study of shrinkage strains in a stereolithography cured acrylic photopolymer resin," "Journal of Materials Processing Technology", Volume 136, Issues 1–3, 10 May 2003, Pages 146-150

[4] Solving Z-axis challenges during stereolithography processes

[5] Determining fabrication orientations for rapid prototyping with Stereolithography apparatus

[6] Shyamasundar, RudrapatnaK. "Feasibility of design in stereolithography," "Foundations of Software Technology and Theoretical Computer Science", Volume 761 Springer, 1993,

[7] D Pham, S Dimov, R Gault, "Part Orientation in Stereolithography," "The International Journal of Advanced Manufacturing Technology", Volume 15, Issue 9, 1999-08-01, Pages 674-682

[8] Specs|Formlabs

[9] http://web.iitd.ac.in/~{}pmpandey/RP_html_pdf/protec_orien.pdf

[10] http://formlabs.com/support/guide/prepare/what-supports-do/

[11] Sambu, S., Y. Chen, and D.W. Rosen, Gometric Tailoring: A Design for Manufacturing Method for Rapid Prototyping and Rapid Tooling. Journal of Mechanical Design, 2004. 126: p. 1-10.

[12] West, A.P., Sambu, S. and Rosen, D.W. (2001), "A process planning method for improving build performance in stereolithography",Computer-Aided Design, Vol. 33, No. 1, pp. 65-80

[13] Lynn-Charney, C.M. and Rosen, D.W. (2000), "Accuracy models and their use in stereolithography process planning",Rapid Prototyping Journal, Vol. 6 No. 2, pp. 77-86

[14] Susman, G.I., Integrating Design and Manufacturing for Competitive Advantage. 1992, New York: Oxford University Press.

[15] A.G.M. Michell "The limits of economy of material in frame-structures", Philosophical Magazine Series 6, Vol. 8, Iss. 47, 1904

[16] The Design of Michell Optimum Structure, NACA

[17] DFM framework for design for additive manufacturing problems

22.6 External links

- DFM framework for design for additive manufacturing problems

- Geometric Tailoring for Rapid Prototyping and Rapid Tooling

- Boothroyd, Geoffrey; Knight, W. A. (Winston Anthony), 1941-; Dewhurst, P. (Peter) (1994), *Product design for manufacture and assembly*, M. Dekker, ISBN 978-0-8247-9176-6

Chapter 23

Digital materialization

Digital materialization (DM) [1] [2] can loosely be defined as two-way direct communication or conversion between matter and information that enables people to exactly describe, monitor, manipulate and create any arbitrary real object. DM is a general paradigm alongside a specified framework that is suitable for computer processing and includes: holistic, coherent, volumetric modeling systems; symbolic languages that are able to handle infinite degrees of freedom and detail in a compact format; and the direct interaction and/or fabrication of any object at any spatial resolution without the need for "lossy" or intermediate formats.

DM systems possess the following attributes:

- realistic - correct spatial mapping of matter to information

- exact - exact language and/or methods for input from and output to matter

- infinite - ability to operate at any scale and define infinite detail

- symbolic - accessible to individuals for design, creation and modification

Such an approach can not only be applied to tangible objects but can include the conversion of things such as light and sound to/from information and matter. Systems to digitally materialize light and sound already largely exist now (e.g. photo editing, audio mixing, etc.) and have been quite effective - but the representation, control and creation of tangible matter is poorly support by computational and digital systems.

Commonplace computer-aided design and manufacturing systems currently represent real objects as "2.5 dimensional" shells. In contrast, DM proposes a deeper understanding and sophisticated manipulation of matter by directly using rigorous mathematics as complete volumetric descriptions of real objects. By utilizing technologies such as Function representation (FRep) it becomes possible to compactly describe and understand the surface and internal structures or properties of an object at an infinite resolution. Thus models can accurately represent matter across all scales making it possible to capture the complexity and quality of natural and real objects and ideally suited for digital fabrication and other kinds of real world interactions. DM surpasses the previous limitations of static disassociated languages and simple human-made objects, to propose systems that are heterogeneous, interacting directly and more naturally with the complex world.[3]

Digital and computer-based languages and processes, unlike the analogue counterparts, can computationally and spatially describe and control matter in an exact, constructive and accessible manner. However, this requires approaches that can handle the complexity of natural objects and materials.

23.1 See also

- Function representation

- Constructive Solid Geometry

- Isosurface

- Solid modeling

- 3D printing

- Additive manufacturing

- Rapid prototyping

- Molecular assembler

- RepRap

23.2 References

[1] T. Vilbrandt, A. Pasko, C. Vilbrandt, Fabricating Nature, Technoetic Arts, Vol. 7, Issue 2, ISSN: 1477965X, Intellect, UK, 2009, pp. 165-174

[2] R. Armstrong, Systems architecture: a new model for sustainability and the built environment using nanotechnology, biotechnology, information technology, and cognitive science with living technology, *Artificial Life*, MIT Press, Vol. 16, No. 1, 2010, pp. 73-87.

[3] T. Vilbrandt, E. Malone, H. Lipson, A. Pasko, Universal Desktop Fabrication, in *Heterogeneous Objects Modelling and Applications, Lecture Notes in Computer Science*, vol. 4889, Springer Verlag, 2008, pp. 259-284

23.3 External links

- Digital Materialization Group

- Computer Aided Design's eXtended Dimensions

- The Self Fab House

- People who have Digital Materialization as a research interest

Chapter 24

Direct metal laser sintering

Direct metal laser sintering (**DMLS**) is an additive manufacturing technique that uses a carbon dioxide laser fired into a magnesium substrate to sinter powdered material (typically metal), aiming the laser automatically at points in space defined by a 3D model, binding the material together to create a solid structure. It is similar to selective laser sintering (SLS); the two are instantiations of the same concept but differ in technical details. Selective laser melting (SLM) uses a comparable concept, but in SLM the material is fully melted rather than sintered, allowing different properties (crystal structure, porosity, and so on). DMLS was developed by the EOS firm of Munich, Germany.[1]

The DMLS process involves use of a 3D CAD model whereby a .stl file is created and sent to the machine's software. A technician works with this 3D model to properly orient the geometry for part building and adds supports structure as appropriate. Once this "build file" has been completed, it is "sliced" into the layer thickness the machine will build in and downloaded to the DMLS machine allowing the build to begin. The DMLS machine uses a high-powered 200 watt Yb-fiber optic laser. Inside the build chamber area, there is a material dispensing platform and a build platform along with a recoater blade used to move new powder over the build platform. The technology fuses metal powder into a solid part by melting it locally using the focused laser beam. Parts are built up additively layer by layer, typically using layers 20 micrometres thick. This process allows for highly complex geometries to be created directly from the 3D CAD data, fully automatically, in hours and without any tooling. DMLS is a net-shape process, producing parts with high accuracy and detail resolution, good surface quality and excellent mechanical properties.

24.1 Benefits

DMLS has many benefits over traditional manufacturing techniques. The ability to quickly produce a unique part is the most obvious because no special tooling is required and parts can be built in a matter of hours. Additionally, DMLS allows for more rigorous testing of prototypes. Since DMLS can use most alloys, prototypes can now be functional hardware made out of the same material as production components.

DMLS is also one of the few additive manufacturing technologies being used in production. Since the components are built layer by layer, it is possible to design internal features and passages that could not be cast or otherwise machined. Complex geometries and assemblies with multiple components can be simplified to fewer parts with a more cost effective assembly. DMLS does not require special tooling like castings, so it is convenient for short production runs.

24.2 Applications

This technology is used to manufacture direct parts for a variety of industries including aerospace, dental, medical and other industries that have small to medium size, highly complex parts and the tooling industry to make direct tooling inserts. With a typical build envelope (e.g. for EOS's EOSINT M280[2]) of 250 x 250 x 325 mm, and the ability to 'grow' multiple parts at one time, DMLS is a very cost and time effective technology. The technology is used both for rapid prototyping, as it decreases development time for new products, and production manufacturing as a cost saving method to simplify assemblies and complex geometries.[3]

The Northwestern Polytechnical University of China is using a similar system to build structural titanium parts for aircraft.[4] An EADS study shows that use of the process would reduce materials and waste in aerospace applications.[5]

On September 5, 2013 Elon Musk tweeted an image of SpaceX's regeneratively-cooled SuperDraco rocket engine chamber emerging from an EOS 3D metal printer, noting that it was composed of the Inconel superalloy.[6] In a surprise move, SpaceX announced in May 2014 that the flight-qualified version of the SuperDraco engine is fully printed, and is the first fully printed rocket engine. Using Inconel, an alloy of nickel and iron, additively-manufactured by direct metal laser sintering, the engine operates at a chamber pressure of 6,900 kilopascals (1,000 psi) at a very high temperature. The engines are contained in a printed protective nacelle, also DMLS-printed, to prevent fault propagation in the event of an engine failure.[7][8][9] The engine completed a full qualification test in May 2014, and is slated to make its first orbital spaceflight in 2015 or 2016.[9]

The ability to 3D print the complex parts was key to achieving the low-mass objective of the engine. According to Elon Musk, "It's a very complex engine, and it was very difficult to form all the cooling channels, the injector head, and the throttling mechanism. Being able to print very high strength advanced alloys ... was crucial to being able to create the SuperDraco engine as it is."[10] The 3D printing process for the SuperDraco engine dramatically reduces lead-time compared to the traditional cast parts, and "has superior strength, ductility, and fracture resistance, with a lower variability in materials properties."[11]

24.3   Constraints

The aspects of size, feature details and surface finish, as well as print through error in the Z axis may be factors that should be considered prior to the use of the technology. However, by planning the build in the machine where most features are built in the x and y axis as the material is laid down, the feature tolerances can be managed well. Surfaces usually have to be polished to achieve mirror or extremely smooth finishes.

For production tooling, material density of a finished part or insert should be addressed prior to use. For example, in injection molding inserts, any surface imperfections will cause imperfections in the plastic part, and the inserts will have to mate with the base of the mold with temperature and surfaces to prevent problems.

Independent of the material system used, the DMLS process leaves a grainy surface finish due to "powder particle size, layer-wise building sequence and [the spreading of the metal powder prior to sintering by the powder distribution mechanism]."[12]

Metallic support structure removal and post processing of the part generated may be a time consuming process and require the use of machining, EDM and/or grinding machines having the same level of accuracy provided by the RP machine.

Laser polishing by means of shallow surface melting of DMLS-produced parts is able to reduce surface roughness by use of a fast-moving laser beam providing "just enough heat energy to cause melting of the surface peaks. The molten mass then flows into the surface valleys by surface tension, gravity and laser pressure, thus diminishing the roughness."[12]

When using rapid prototyping machines, .stl files, which do not include anything but raw mesh data in binary (generated from Solid Works, CATIA, or other major CAD programs) need further conversion to .cli & .sli files (the format required for non stereolithography machines).[13] Software converts .stl file to .sli files, as with the rest of the process, there can be costs associated with this step.

24.4   Materials

Currently available alloys used in the process include 17-4 and 15-5 stainless steel, maraging steel, cobalt chromium, inconel 625 and 718, and titanium Ti6Al4V.[14]

24.5   See also

- List of notable 3D printed weapons and parts
- 3D printing
- Additive manufacturing
- Desktop manufacturing
- Digital fabricator
- Direct digital manufacturing
- Fab lab
- Fused deposition modeling
- Instant manufacturing, also known as "direct manufacturing" or "on-demand manufacturing"
- Rapid manufacturing
- Rapid prototyping
- RepRap Project
- Solid freeform fabrication
- Stereolithography
- Laser engineered net shaping
- Laser sintering of gold

24.6   References

[1] "How Direct Metal Laser Sintering Works". THRE3D.com. Retrieved 3 February 2014.

[2] http://ip-saas-eos-cms.s3.amazonaws.com/public/e1dc925774b24d9f/55e7f647441dc9e8fdaf944d18416bdb/systemdatasheet_M280_n.pdf

[3] Additive Companies Run Production Parts

[4] "China commercializes 3D printing in aviation."

[5] "EADS Innovation Works Finds 3D Printing Reduces CO2 by 40%". *3dprintinginsider.com*. Mediabistro Inc. Retrieved 7 November 2013.

[6] "Twitter". Mobile.twitter.com. Retrieved 2014-08-21.

[7] Norris, Guy (2014-05-30). "SpaceX Unveils 'Step Change' Dragon 'V2'". *Aviation Week*. Retrieved 2014-05-30.

[8] Kramer, Miriam (2014-05-30). "SpaceX Unveils Dragon V2 Spaceship, a Manned Space Taxi for Astronauts — Meet Dragon V2: SpaceX's Manned Space Taxi for Astronaut Trips". *space.com*. Retrieved 2014-05-30.

[9] Bergin, Chris (2014-05-30). "SpaceX lifts the lid on the Dragon V2 crew spacecraft". *NASAspaceflight.com*. Retrieved 2015-03-06.

[10] Foust, Jeff (2014-05-30). "SpaceX unveils its "21st century spaceship"". *NewSpace Journal*. Retrieved 2015-03-06.

[11] "SpaceX Launches 3D-Printed Part to Space, Creates Printed Engine Chamber for Crewed Spaceflight". SpaceX. Retrieved 2015-03-06. *Compared with a traditionally cast part, a printed [part] has superior strength, ductility, and fracture resistance, with a lower variability in materials properties. ... The chamber is regeneratively cooled and printed in Inconel, a high performance superalloy. Printing the chamber resulted in an order of magnitude reduction in lead-time compared with traditional machining – the path from the initial concept to the first hotfire was just over three months. During the hotfire test, ... the SuperDraco engine was fired in both a launch escape profile and a landing burn profile, successfully throttling between 20% and 100% thrust levels. To date the chamber has been fired more than 80 times, with more than 300 seconds of hot fire.*

[12] "Surface Roughness Enhancement of Indirect-SLS Metal Parts by Laser Surface Polishing" (PDF). University of Texas at Austin. 2001. Retrieved 2015-10-12.

[13] http://knowledge.stereolithography.com/activekb/questions/74/STL+File+Conversion

[14] http://www.eos.info/material-m

24.7 External links

- Rapid Manufacturing's Role in the Factory of the Future

- Direct metal laser sintering, video (2:34).

- [1]

[1] Thu, 02/06/2014 - 10:41am (2014-02-06). "The Laser-Sintering Effect". Rdmag.com. Retrieved 2014-08-21.

Chapter 25

Distributed manufacturing

Distributed manufacturing also known as **distributed production** and **local manufacturing** is a form of decentralized manufacturing practiced by enterprises using a network of geographically dispersed manufacturing facilities that are coordinated using information technology. It can also refer to local manufacture via the historic cottage industry model, or manufacturing that takes place in the homes of consumers.

25.1   Enterprise

The primary attribute of distributed manufacturing is the ability to create value at geographically dispersed locations via manufacturing. For example, shipping costs are minimized when products are built geographically close to their intended markets. Also, products manufactured in a number of small facilities distributed over a wide area can be customized with details adapted to individual or regional tastes. Manufacturing components in different physical locations and then managing the supply chain to bring them together for final assembly of a product is also considered a form of distributed manufacturing.[1][2] Digital networks combined with additive manufacturing allow companies a decentralized and geographically independent distributed production (Cloud Producing).[3]

25.2   Consumer

Within the maker movement and DIY culture, small scale production by consumers often using peer to peer resources is being referred to as distributed manufacturing. Consumers download digital designs from an open design repository website like Youmagine or Thingiverse and produce a product at home for low costs with an open-source 3-D printer such as the RepRap.[4][5] Distributed manufacturing with distributed generation using solar photovoltaic cells and 3-D printers has been proposed as a means for off-grid rural area residents to manufacture themselves out of poverty.[6]

An example of such an application are spectacles. As Gwamuri et al. point out that while it is "still not yet feasible to print the lenses (the most critical component of the eyeglasses)" and in current prototypes "only the frames and syringe are printed" and that "aesthetics is another challenge" the "primary cost of the glasses could be reduced to about one dollar for a highly customized/individualized design, which could be printed on site in under an hour" (presumably excluding the lenses) and "it seems clear that other products could benefit from the same approach and that distributed manufacturing can assist in sustainable development, particularly in isolated rural regions".[7]

Initial life cycle analysis indicates that distributed production can have a smaller impact on the environment than conventional manufacturing and shipping because of reductions in transportation embodied energy.[8][9] There are now several types of open-source solar-powered 3-D printers,[10] which can be used for production in off grid locations.[11]

25.3 Social change

Some[12][13][14] call attention to the conjunction of Commons-based peer production with distributed manufacturing techniques. The self-reinforced fantasy of a system of eternal growth can be overcome with the development of economies of scope, and here, the civil society can play an important role contributing to the raising of the whole productive structure to a higher plateau of more sustainable and customised productivity.[12] Further, it is true that many issues, problems and threats rise due to the large democratisation of the means of production, and especially regarding the physical ones.[12] For instance, the recyclability of advanced nanomaterials is still questioned; weapons manufacturing could become easier; not to mention the implications on counterfeiting[15] and on "intellectual property".[16] It might be maintained that in contrast to the industrial paradigm whose competitive dynamics were about economies of scale, Commons-based peer production and distributed manufacturing could develop economies of scope. While the advantages of scale rest on cheap global transportation, the economies of scope share infrastructure costs (intangible and tangible productive resources), taking advantage of the capabilities of the fabrication tools.[12] And following Neil Gershenfeld[17] in that "some of the least developed parts of the world need some of the most advanced technologies", Commons-based peer production and distributed manufacturing may offer the necessary tools for thinking globally but act locally in response to certain problems and needs.

25.4 See also

- Commons-based peer production

- Fablab

- Peer production

- RepRap

- Recyclebot

25.5 References

[1] Chrisman, Ray. "Enhancement of Distributed Manufacturing using expanded Process Intensification Concepts" (PDF). University of Washington. Retrieved 7 May 2013.

[2] Hermann Kühnle (2010). *Distributed Manufacturing: Paradigm, Concepts, Solutions and Examples*. Springer. ISBN 978-1-84882-707-3. Retrieved 7 May 2013.

[3] Felix Bopp (2010). *Future Business Models by Additive Manufacturing*. Verlag. ISBN 3836685086. Retrieved 4 July 2014.

[4] Sells, Ed, Zach Smith, Sebastien Bailard, Adrian Bowyer, and Vik Olliver. "Reprap: the replicating rapid prototyper: maximizing customizability by breeding the means of production." HANDBOOK OF RESEARCH IN MASS CUSTOMIZATION AND PERSONALIZATION, (2010).

[5] Jones, R., Haufe, P., Sells, E., Iravani, P., Olliver, V., Palmer, C., & Bowyer, A. (2011). Reprap??? the replicating rapid prototyper. Robotica, 29(1), 177-191.

[6] Pearce, J. M., Blair, C. M., Laciak, K. J., Andrews, R., Nosrat, A., & Zelenika-Zovko, I. (2010). 3-D printing of open source appropriate technologies for self-directed sustainable development. Journal of Sustainable Development, 3(4), p17.

[7] J. Gwamuri, B. T. Wittbrodt, N. C. Anzalone, J.M. Pearce. Reversing the Trend of Large Scale and Centralization in Manufacturing: The Case of Distributed Manufacturing of Customizable 3-D-Printable Self-Adjustable Glasses, *Challenges in Sustainability* 2(1), pp. 30-40 (2014). DOI: 10.12924/cis2014.02010030

[8] M. Kreiger, G. C. Anzalone, M. L. Mulder, A. Glover and J. M Pearce (2013). Distributed Recycling of Post-Consumer Plastic Waste in Rural Areas. MRS Online Proceedings Library, 1492, mrsf12-1492-g04-06 doi:10.1557/opl.2013.258. open access

[9] Megan Kreiger and Joshua M. Pearce (2013). Environmental Life Cycle Analysis of Distributed 3-D Printing and Conventional Manufacturing of Polymer Products, *ACS Sustainable Chemistry & Engineering*, DOI: 10.1021/sc400093k Open access.

[10] Debbie L. King, Adegboyega Babasola, Joseph Rozario, and Joshua M. Pearce, "Mobile Open-Source Solar-Powered 3-D Printers for Distributed Manufacturing in Off-Grid Communities," Challenges in Sustainability 2(1), 18-27 (2014).open access

[11] D.J. Pangburn, How 3D Printers Are Boosting Off-The-Grid, Underdeveloped Communities - MotherBoard available at http://motherboard.vice.com/read/how-3d-printers-are-boosting-off-the-grid-underdeveloped-communities Nov. 7, 2014.

[12] Kostakis, V.; Bauwens, M. (2014): *Network Society and Future Scenarios for a Collaborative Economy*. Basingstoke, UK: Palgrave Macmillan. (wiki)

[13] Kostakis, V.; Papachristou, M. (2014): *Commons-based peer production and digital fabrication: The case of a RepRap-based, Lego-built 3D printing-milling machine*. In: Telematics and Informatics, 31(3), 434 - 443

[14] Kostakis, V; Fountouklis, M; Drechsler, W. (2013): *Peer Production and Desktop Manufacturing: The Case of the Helix-T Wind Turbine Project*. . In: Science, Technology & Human Values, 38(6), 773 - 800.

[15] Campbell, Thomas, Christopher Williams, Olga Ivanova, and Banning Garrett. (2011): *Could 3D Printing Change the World? Technologies, Potential, and Implications of Additive Manufacturing*. Washington: Atlantic Council of the United States

[16] Bradshaw, Simon, Adrian Bowyer, and Patrick Haufe (2010): *The Intellectual Property Implications of Low-Cost 3D Printing*. In: SCRIPTed 7

[17] Gershenfeld, Neil (2007): *FAB: The Coming Revolution on your Desktop: From Personal Computers to Personal Fabrication*. Cambridge: Basic Books, p. 13-14

Chapter 26

Electron beam additive manufacturing

For other uses, see electron beam furnace.

Electron beam additive manufacturing is a type of additive manufacturing, or 3D printing, for metal parts. Metal powder or wire is welded together using an electron beam as the heat source.

26.1 Metal powder based systems

Metal powders can be sintered into a solid mass using an electron beam as the heat source. This is used as a 3D printing technique, similar to selective laser sintering. This is sometimes called "electron beam melting". EBM technology manufactures parts by melting metal powder layer by layer with an electron beam in a high vacuum. In contrast to sintering techniques, both EBM and SLM achieve full melting of the metal powder. [1]

This solid freeform fabrication method produces fully dense metal parts directly from metal powder with characteristics of the target material. The EBM machine reads data from a 3D CAD model and lays down successive layers of powdered material. These layers are melted together utilizing a computer controlled electron beam. In this way it builds up the parts. The process takes place under vacuum, which makes it suited to manufacture parts in reactive materials with a high affinity for oxygen, e.g. titanium.[2] The process is known to operate at higher temperatures (up to 1000 °C), which can lead to differences in phase formation though solidification and solid state phase transformation.[3]

The powder feedstock is typically pre-alloyed, as opposed to a mixture. That aspect allows classification of EBM with selective laser melting (SLM) where competing technologies like SLS and DMLS require thermal treatment after fabrication. Compared to SLM and DMLS, EBM has a generally superior build rate because of its higher energy density and scanning method.

26.1.1 Research Developments

Recent work has been published by ORNL, demonstrating the use of EBM technology to control local crystallographic grain orientations in Inconel.[4] Other notable developments have focused on the development of process parameters to produced parts out of alloys such as copper,[5] niobium,[6] Al 2024,[7] bulk metallic glass,[8] stainless steel, and titanium aluminide. Currently commercial materials for EBM include commercially pure Titanium, Ti-6Al-4V,[9] CoCr, Inconel 718,[10] and Inconel 625.[11]

26.2 Metal wire based systems

Another approach is to use an electron beam to melt welding wire onto a surface to build up a part.[12] This is similar to the common 3D printing process of fused deposition modeling, but with metal, rather than plastics. With this process, an electron beam gun provides the energy source used for melting metallic feedstock, which is typically wire. The electron beam is a highly efficient power source that can be both precisely focused and deflected using electromagnetic coils at rates well into thousands of hertz. Typical electron beam welding systems have high power availability, with 30- and 42-kilowatt systems being most common. A major advantage of using metallic components with electron beams is that the process is conducted within a high vacuum environment of 1x10-4 Torr. or greater, providing a contamination-free work zone that does not require the use of additional inert gasses commonly used with laser and arc based processes. With EBDM, feedstock material is fed into a molten pool created by the electron beam. Through the use of computer numeric controls (CNC), the molten pool is moved about on a substrate plate, adding material just where it is needed to produce the near net shape. This process is repeated in a layer-by-layer fashion, until the desired 3D shape is produced.

Depending on the part being manufactured, deposition rates can range up to 200 cubic inches per hour. With a light alloy, such as titanium, this translates to a real-time deposition rate of 40 pounds per hour. A wide range of engineering alloys are compatible with the EBDM process and are readily available in the form of welding wire from an existing supply base. These include, but are not limited to, stainless steels, cobalt alloys, nickel alloys, copper nickel alloys, tantalum, titanium alloys as well as many other high-value materials.

26.3 Market

Titanium alloys are widely used with this technology which makes it a suitable choice for the medical implant market.

CE-certified acetabular cups are in series production with EBM since 2007 by two European orthopedic implant manufacturers, Adler Ortho and Lima Corporate.

The U.S. implant manufacturer Exactech has also received FDA clearance for an acetabular cup manufactured with the EBM technology.

Aerospace and other highly demanding mechanical applications are also targeted.

The EBM process has been developed for manufacturing parts in gamma titanium aluminide, and is currently being developed by Avio S.p.A. and General Electric Aviation for the production of turbine blades in γ-TiAl for gas-turbine engines.[13]

26.4 See also

- 3D printing

- Electron beam technology

- Electron beam welding

26.5 References

[1] ASTM International, ASTM F2792-12a, http://www.astm.org/Standards/F2792.htm

[2] "Electron Beam Melting". Thre3d.com. Retrieved 28 January 2014.

[3] Sames et al., "Thermal effects on microstructural heterogeneity of Inconel 718 materials fabricated by electron beam melting", http://dx.doi.org/10.1557/jmr.2014.140

[4] "ORNL research reveals unique capabilities of 3-D printing", http://www.ornl.gov/ornl/news/news-releases/2014/ornl-research-reveals-unique-cap

[5] Frigola et al., "Fabricating Copper Components with Electron Beam Melting", http://www.asminternational.org/documents/10192/19735983/amp17207p20.pdf/2a87d5ae-86ec-4f27-bdd1-f74af1a2f523

[6] Martinez et al., "Microstructures of Niobium Components Fabricated by Electron Beam Melting", http://link.springer.com/article/10.1007%2Fs13632-013-0073-9

[7] Mahale, Ph.D. Thesis, NCSU 2009, http://adsabs.harvard.edu/abs/2009PhDT.......262M

[8] http://www.sciencedaily.com/releases/2012/11/121119114157.htm

[9] http://www.arcam.com/technology/electron-beam-melting/materials/

[10] Sames et al., "Effect of Process Control and Powder Quality on Inconel 718 Produced Using Electron Beam Melting", http://www.programmaster.org/PM/PM.nsf/ViewSessionSheets?OpenAgent&ParentUNID=733A03EC3B3DE9B485257C5A00064CE6

[11] Murr et al., "Fabrication of Metal and Alloy Components by Additive Manufacturing: Examples of 3D Materials Science", http://www.jmrt.com.br/en/fabrication-of-metal-and-alloy/articulo/90195169/

[12] "Video: Electron Beam Direct Manufacturing : Modern Machine Shop". Mmsonline.com. Retrieved 10 October 2013.

[13] http://3dprint.com/12262/ge-ebm-3d-printing/

26.6 Further reading

- Manufacturing Engineering and Technology Fifth Edition. Serope Kalpakjian.

26.7 External links

- Watch and Learn about electron beam melting
- Engineer's Handbook
- Automotive DesignLine
- Arcam process presentation (pdf)
- Direct Manufacturing: ARCAM, Video Describing EBM Process

Chapter 27

Electron beam freeform fabrication

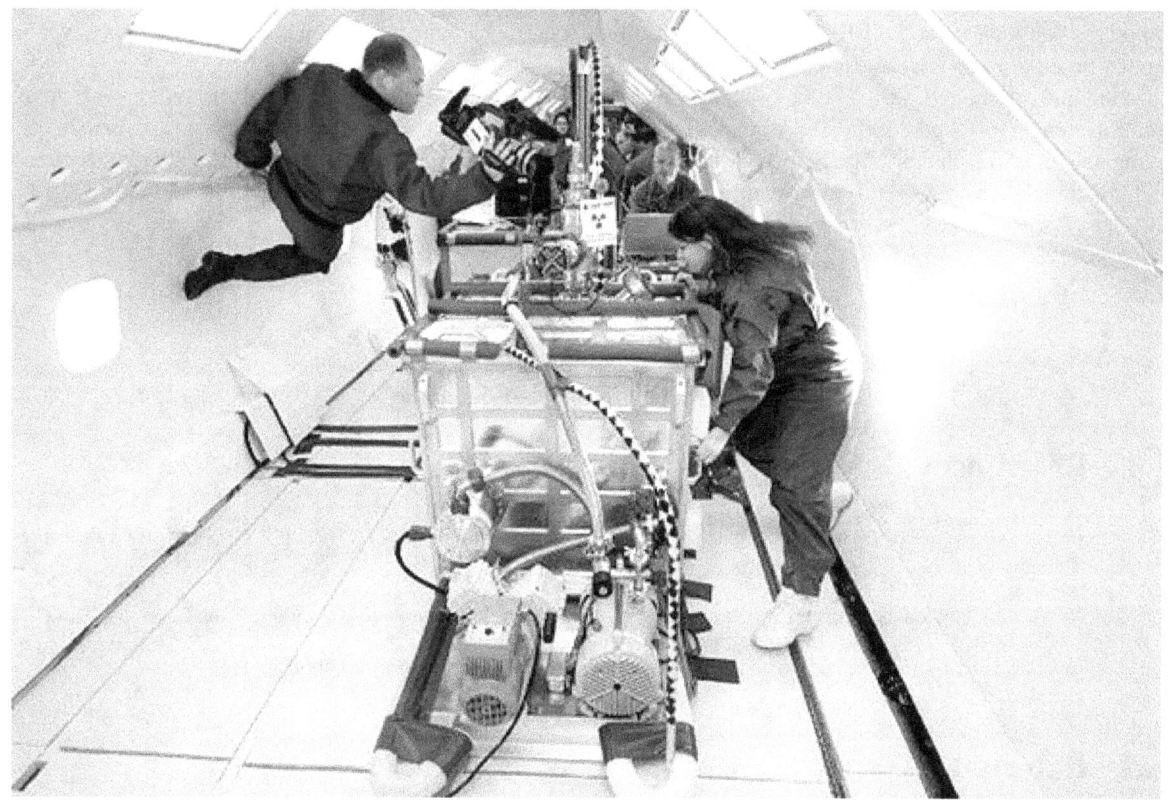

NASA engineers test the EBF³ system during a parabolic flight in 2007.

Electron Beam Freeform Fabrication (**EBF³**) is an additive manufacturing process that builds near-net-shape parts requiring less raw material and finish machining than traditional manufacturing methods. It uses a focused electron beam in a vacuum environment to create a molten pool on a metallic substrate.

27.1 History

NASA Langley Research Center (LaRC) originated (EBF³) technology development. The Additive Manufacturing Process was primarily developed and engineered by Karen Taminger, material research engineer for NASA LaRC. EBF³ is a NASA-patented additive manufacturing process designed to build near-net-shape parts requiring less raw material and finish machining than traditional manufacturing methods. EBF^3 is a process by which NASA plans to build metal parts in zero gravity environments; this layer-additive process uses an electron beam, and a solid wire feedstock to fabricate metallic structures. The process efficiencies of the electron beam and the feedstock make the EBF^3 process appropriate for in-space use. Since 2000, a Team of Researchers at the NASA LaRC have led the fundamental research and development of this technique for additive manufacturing; which is for metallic aerospace structures. Additive manufacturing encompasses processes in which parts are built by successively adding material rather than by cutting or grinding it away as in conventional machining. Additive manufacturing is an outgrowth of rapid prototyping techniques such as stereolithography, first developed for non-structural plastic parts over thirty years ago.[1]

27.2 Process

The operational concept of EBF^3 is to build a near-net-shape metal part directly from a Computer Aided Design (CAD) file. Current computer-aided machining practices start with a CAD model and use a post-processor to write the machining instructions (G-code) defining the cutting tool paths needed to make the part. EBF^3 uses a similar process, starting with a CAD model, numerically slicing it into layers, then using a post-processor to write the G-code defining the deposition path and process parameters for the EBF^3 equipment.[2] It uses a focused electron beam in a vacuum environment to create a molten pool on a metallic substrate. The beam is translated with respect to the surface of the substrate while metal wire is fed into the molten pool. The deposit solidifies immediately after the electron beam has passed, having sufficient structural strength to support itself. The sequence is repeated in a layer-additive manner to produce a near-net-shape part needing only finish machining. The EBF^3 process is scalable for components from fractions of an inch to tens of feet in size, limited mainly by the size of the vacuum chamber and amount of wire feedstock available.[3]

27.3 See also

- Electron beam additive manufacturing

27.4 References

[1] http://ntrs.nasa.gov/archive/nasa/casi.ntrs.nasa.gov/20080013538_2008013396.pdf Electron Beam Freeform Fabrication for Cost Effective Near-Net Shape

[2] http://www.nasa.gov/topics/aeronautics/features/electron_beam.html From Nothing, Something: One Layer at a Time

[3] http://www.techbriefs.com/content/view/478/34/Portable Electron-Beam Free-Form Fabrication System

27.5 External links

- Video: EBF3 – Electron Beam Free Form Fabrication

- Electron Beam Freeform Fabrication for Cost Effective Near-Net Shape

- From Nothing, Something: One Layer at a Time

- Electron-Beam Free-Form Fabrication System

- Device like 'Star Trek' replicator is in the works

Chapter 28

Fab@Home

Fab@Home was the first multi-material 3D printer available to the public, and one of the first two open-source DIY 3D printers (the other one being the RepRap). Up until 2005, all 3D printers were industrial scale, expensive and proprietary. The high cost and closed nature of the 3D printing industry at the time limited the accessibility of the technology to the masses, the range of materials that could be used and the level of exploration that could be done by end-users. The goal of the Fab@Home project was to change this situation by creating a versatile, low-cost, open and "hackable" printer to accelerate technology innovation and its migration into the consumer and Maker space.

Since its open-source release in 2006,[1] hundreds of Fab@Home 3D printers were built across the world,[2] and its design elements could be found in many later DIY printers, most notably the Makerbot Replicator. The printer's multiple syringe-based deposition method allowed for some of the first multi-material prints including direct fabrication of active batteries, actuators, and sensors, as well as esoteric materials for bioprinting and food printing.[3] The project was closed in 2012 when it was clear that the project's goal was achieved and distribution of DIY and consumer printers outpaced the sales of industrial printers for the first time.[3]

28.1   History

The project was led by students at Cornell University's department of Mechanical & Aerospace Engineering. The effort was inspired by the history of the Altair 8800, one of the first DIY home computer kits released in 1975. The Altair 8800 is largely credited with triggering the home computing revolution and the transition from the industrial mainframe to the consumer desktop, by making a low-cost, open and "hackable" computer accessible to home enthusiasts for the first time. The goal of the Fab@Home project was to accomplish a similar effect in the 3D printing space. The project was one of the first larger scale cases that applied the open source development model to physical devices, a process that later became known as Open Source Hardware.

Early versions of the device were produced and refined in the lab. The first official release of the Fab@Home model 1 coincided with a presentation at the Solid Freeform Fabrication conference in 2006.[1] After its first release, undergraduate and graduate students at Cornell and other locations joined the team and developed an improved version, later released as the Fab@Home Model 2.[4] The main improvements included easier assembly, no soldering, and fewer parts. The team then expanded and developed the model 3. One important variation of Fab@Home was the Fab@School project, which explored the use of 3D printers suitable for classroom use in elementary grades. Fab@School printers could print with benign materials such as Play-Doh and included a safety enclosure.

The project received wide media attention in its initial years, bringing 3D printing from a relatively obscure technology to broader attention. Notable recognition was the Popular Mechanics Breakthrough award[5] and the Rapid Prototyping Journal best paper of the year award.

28.2 Technical abilities

The Fab@Home is a syringe-based deposition system. An X-Y-Z gantry system moves a syringe pump across a 20×20×20 cm (7.87x7.87x7.87 inch) build volume at a maximum speed of 10 mm/s and resolution of 25 μm. Multiple syringes can be controlled independently to deposit material through syringe tips. The syringe displacement could be controlled with microliter precision.

The first version of Fab@Home print head had two syringes; later version had more syringes, up to one print head that had eight syringes that could be used simultaneously.

One of the key advantages of using a syringe based deposition method was that a broad range of materials could be deposited, essentially any liquid, paste, gel or slurry that could be squeezed through a syringe tip. That versatility allowed going beyond printing just in thermoplastics, as did the RepRap and most consumer-scale 3D printers that followed. The range of materials that could be printed with a Fab@Home included hard materials such as epoxy, elastomers such as silicone, biological materials such as cell-seeded hydrogels, food materials such as chocolate, cookie dough and cheese,

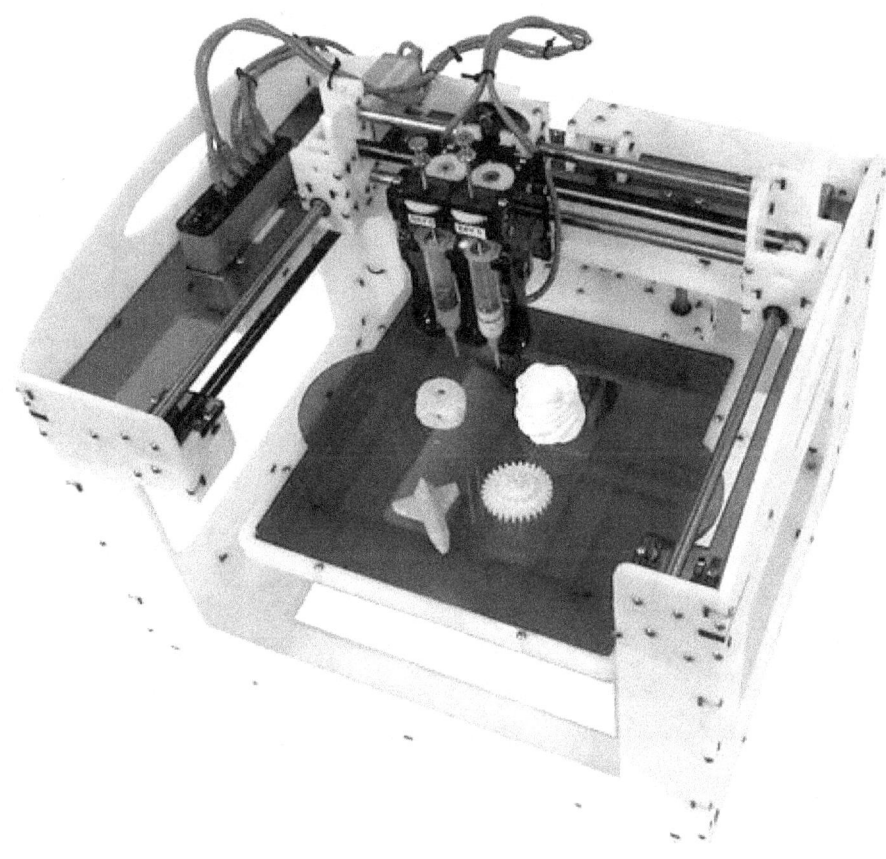

Fab@Home Model 2 (2009)

engineering materials such as stainless steel (later sintered in an oven), and active materials such as conductive wires and magnets. The stated technical goal of the project was to print a complete active system, going beyond printing just passive parts. The project succeeded in printing active devices such as batteries, actuators, sensors, and even a working telegraph machine.

28.3 Project members

- Founders: Evan Malone and Hod Lipson

- Project leads: Evan Malone (2005-2009), Daniel Cohen (2010), Jeffery Lipton (2011-2012)

- Project team members (in no particular order): Dan Periard, Max Lobovsky, James Smith, Michael Heinz, Warren Parad, Garrett Bernstien, Tianyou Li, Justin Quartiere, Daniel Sheiner, Kamaal Washington, Abdul-Aziz Umaru, Rian Masanoff, Justin Granstein, Jordan Whitney, Scott Lichtenthal, Karl Gluck

28.4 See also

- The RepRap project

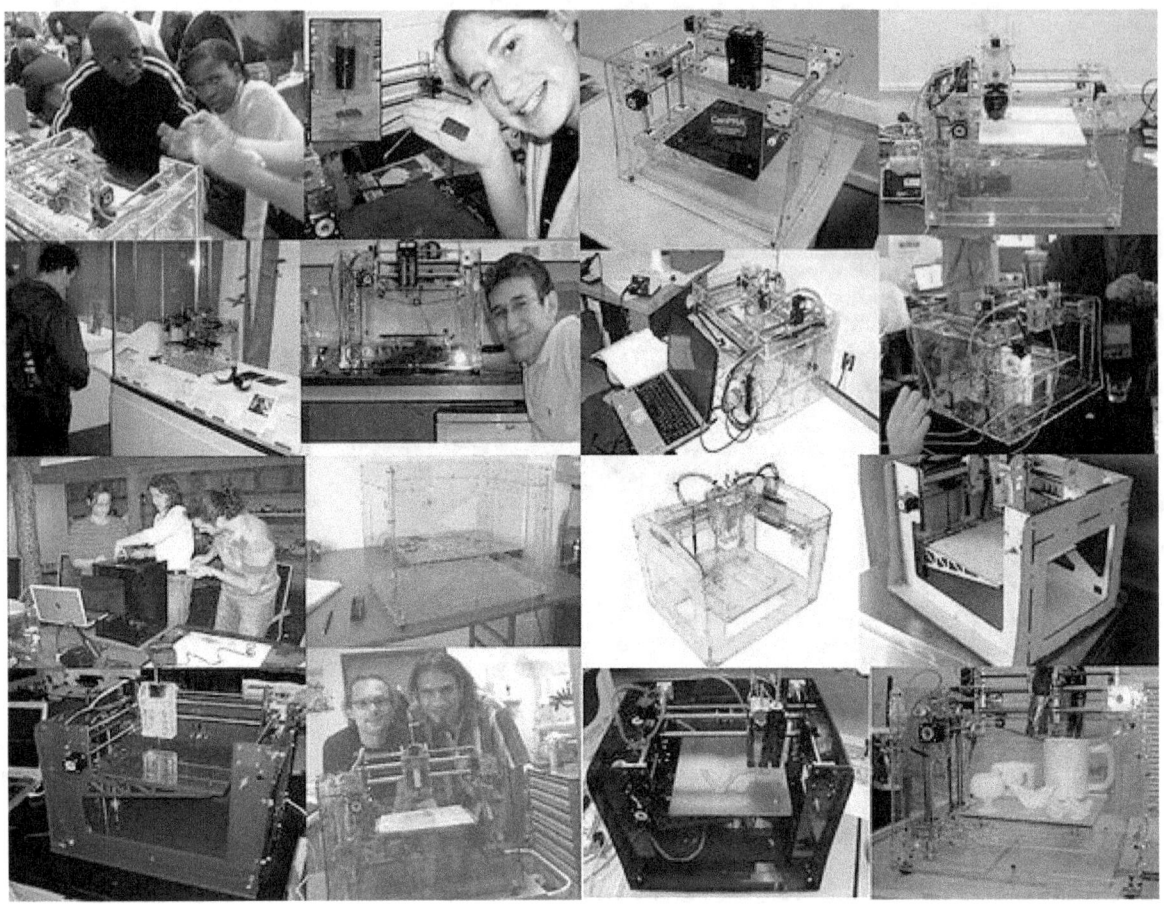

Fab@Home in use

28.5 References

[1] Malone E., Lipson H., (2006) Fab@Home: The Personal Desktop Fabricator Kit, Proceedings of the 17th Solid Freeform Fabrication Symposium, Austin TX, Aug 2006

[2] Additive Manufacturing and 3D Printing State of the Industry, 2012 Annual Worldwide Progress Report, ISBN 0-9754429-8-8

[3] Hod Lipson and Melba Kurman, Fabricated: The new World of 3D printing, Wiley Press, 2013

[4] Lipton, J. Cohen,D., Heinz,M., Lobovsky, M., Parad,W., Bernstien, G., Li,T., Quartiere,J., Washington,K., Umaru,A., Masanoff,R., Granstein, J., Whitney,J., Lipson,H., (2009) "Fab@Home Model 2: Towards Ubiquitous Personal Fabrication Devices" Solid Freeform Fabrication Symposium (SFF'09), Aug 3-5 2009, Austin, TX, USA.

[5] 2007 Popular Mechanics Breakthrough Award, Fab at Home, Open-Source 3D Printer, Lets Users Make Anything

28.6 External links

- fab@home

Chapter 29

Fab lab

A **fab lab** (*fabrication laboratory*) is a small-scale workshop offering (personal) digital fabrication.[1][2]

A fab lab is generally equipped with an array of flexible computer controlled tools that cover several different length scales and various materials, with the aim to make "almost anything".[3] This includes technology-enabled products generally perceived as limited to mass production.

While fab labs have yet to compete with mass production and its associated economies of scale in fabricating widely distributed products, they have already shown the potential to empower individuals to create smart devices for themselves. These devices can be tailored to local or personal needs in ways that are not practical or economical using mass production.

29.1 History

The fab lab program was initiated to broadly explore how the content of information relates to its physical representation and how an under-served community can be powered by technology at the grassroots level.[4] The program began as a collaboration between the Grassroots Invention Group and the Center for Bits and Atoms at the Media Lab in the Massachusetts Institute of Technology with a grant from the National Science Foundation (Washington, D.C.) in 2001.[5]

While the Grassroots Invention Group is no longer in the Media Lab, The Center for Bits and Atoms consortium is still actively involved in continuing research in areas related to description and fabrication but does not operate or maintain any of the labs worldwide (with the excmobile fab lab). The fab lab concept also grew out of a popular class at MIT (MAS.863) named "How To Make (Almost) Anything". The class is still offered in the fall semesters.

29.2 Popular equipment and projects

Flexible manufacturing equipment within a fab lab can include:

- Mainly, a rapid prototyper: typically a 3D printer of plastic or plaster parts

- 3-axis CNC machines: 3 or more axes, computer-controlled subtractive milling or turning machines

- Printed circuit board milling: two-dimensional, high precision milling to create circuit traces in pre-clad copper boards

- Microprocessor and digital electronics design, assembly, and test stations

- Cutters, for sheet material: laser cutter, plasma cutter, water jet cutter, knife cutter

Fab Lab Logo

29.2.1 FabFi

Main article: FabFi

One of the larger projects undertaken by fab labs include free community FabFi wireless networks (in Afghanistan, Kenya

Amsterdam Fab Lab at The Waag Society

and the US). The first city-scale FabFi network, set up in Afghanistan, has remained in place and active for three years under community supervision and with no special maintenance. The network in Kenya, (Based in the University of Nairobi (UoN)) building on that experience, started to experiment with controlling service quality and providing added services for a fee to make the network cost-neutral. plc (programmable logic controls)

29.3   List of labs

MIT maintained a listing of all official Fab Labs, worldwide, until 2014. Nowadays listing of all official Fab Labs maintained by community using site fablabs.io. As of September 2015 there were 107 Fab labs in the US and Canada, and 270 in Europe (565 in the world in total).[6] Currently there are Fab Labs on every continent except Antarctica.

29.4   See also

- TechShop

- 3D printing#3D printing services

- Hackerspace

- Open design

- Open hardware

- Open Source Ecology

- RepRap - The RepRap project aims to produce a free and open source software (FOSS) 3D printer.

29.5 References

[1] Menichinelli, Massimo. "Business Models for Fab Labs".

[2] Troxler, Peter (2011). "Libraries of the Peer Production Era". In van Abel, Bas; Evers, Lucas; Klaassen, Roel; Troxler, Peter. *Open Design Now. Why Design Cannot Remain Exclusive*. Bis Publishers. ISBN 978-90-6369-259-9.

[3] Gershenfeld, Neil A. (2005). *Fab: the coming revolution on your desktop—from personal computers to personal fabrication*. New York: Basic Books. ISBN 0-465-02745-8.

[4] Mikhak, Bakhtiar; "development by design" (dyd02) (2002). "Fab Lab: an alternate model of ICT for development" (PDF). Bangalore ThinkCycle. Retrieved 6 July 2013.

[5] "Fab Central - Fab Lab - IaaC". Retrieved 31 January 2014.

[6] Fab Lab List

29.6 Further reading

- Gershenfeld, Neil A. (2005). *Fab: the coming revolution on your desktop—from personal computers to personal fabrication*. New York: Basic Books. ISBN 0-465-02745-8.

- Walter-Herrmann, Julia & Bueching, Corinne (2013)(eds.) *FabLab - Of Machines, Makers and Inventors*. Bielefeld, Germany: Transcript. ISBN 978-3-8376-2382-6

Chapter 30

Filaments evaluation protocol

The "**filaments evaluation protocol**" (FEP)[1] is a protocol developed to establish the characteristics and qualities of materials used in FDM/FFF 3D printing.

30.1   Purpose

The protocol, designed in 2014 by Dogma Solutions, is not applied to measure the properties of materials, but is instead designed to determine their suitability to specific uses and applications.

For this reason, the evaluations performed during the test does not provide quantitative outputs, but rather indicate, for each tested parameter, a degree of judgment on a unipolar odd Likert scale (by now known as "Degree of Likert Rating" or "DLR"):

1. Strongly discouraged / highly unfit / very difficult to use / test totally failed

2. Discouraged / unsuitable / difficult to use / test partially failed

3. Neither recommended nor discouraged. There are not particular defects nor significant points in favor. The test is successful but the result is not convincing or either qualitatively relevant.

4. Recommended / suitable / easy to use / tested successfully within realistic expectations

5. Strongly recommended / highly suitable / easy to use / test perfectly succeeded up to the best (or even over) expectations

30.2   References

[1] "F.E.P.". *dogmasolutions.com*. External link in |work= (help)

Chapter 31

Fused deposition modeling

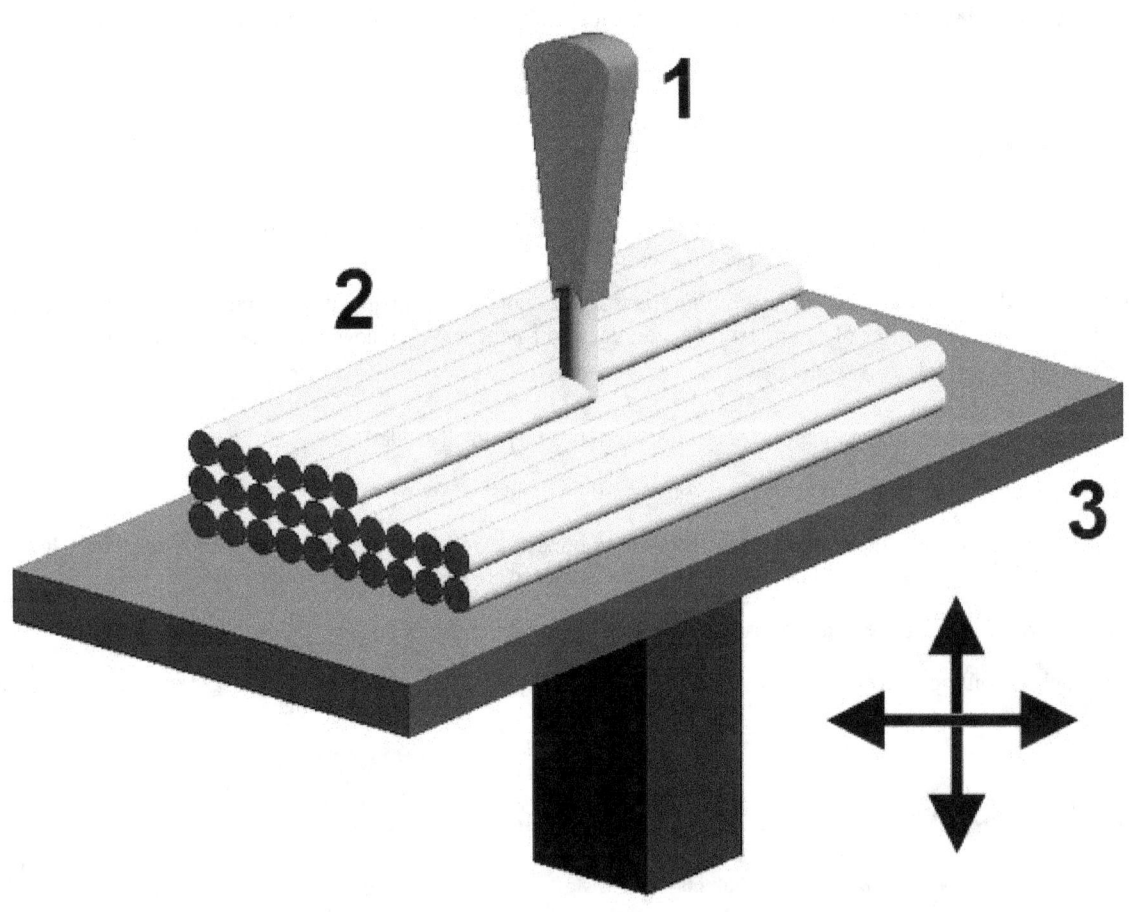

Fused deposition modelling: 1 – nozzle ejecting molten material, 2 – deposited material (modeled part), 3 – controlled movable table

Fused deposition modeling (**FDM**) is an additive manufacturing technology commonly used for modeling, prototyping, and production applications. It is one of the techniques used for 3D printing.

FDM works on an "additive" principle by laying down material in layers; a plastic filament or metal wire is unwound from a coil and supplies material to produce a part.

The technology was developed by S. Scott Crump in the late 1980s and was commercialized in 1990.[1] The term *fused deposition modeling* and its abbreviation to *FDM* are trademarked by Stratasys Inc. The exactly equivalent term, ***fused***

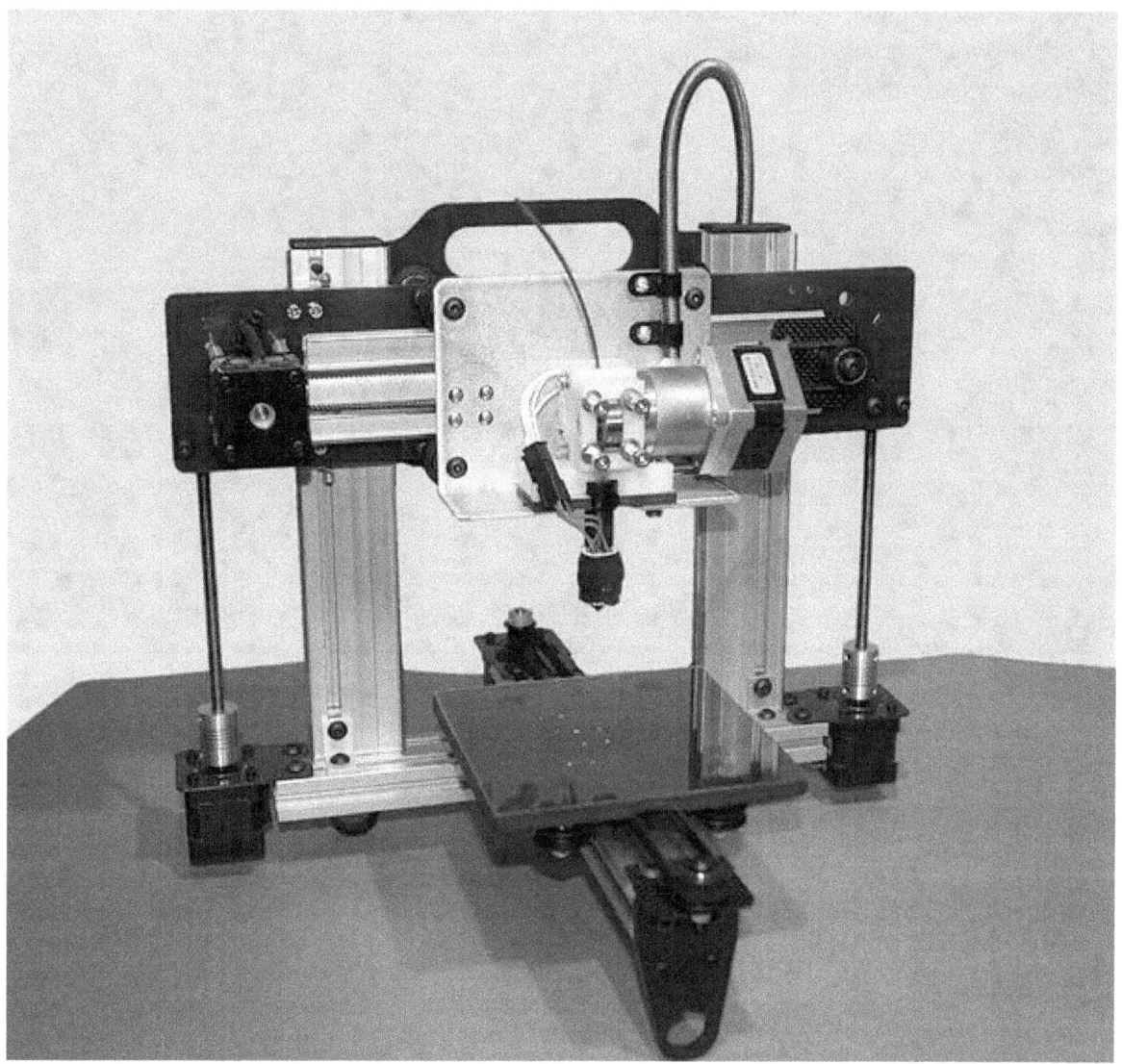

An ORDbot Quantum 3D printer.

filament fabrication (***FFF***), was coined by the members of the RepRap project to give a phrase that would be legally unconstrained in its use. It is also sometimes called Plastic Jet Printing (PJP).

31.1   History

Fused deposition modeling (FDM) was developed by S. Scott Crump in the late 1980s and was commercialized in 1990 by Stratasys.[2] With the expiration of the patent on this technology there is now a large open-source development community, as well as commercial and DIY variants, which utilize this type of 3D printer. This has led to two orders of magnitude price drop since this technology's creation.

Timelapse video of a hyperboloid object (designed by George W. Hart) made of PLA using a RepRap "Prusa Mendel" 3D printer for molten polymer deposition.

31.2 Process

FDM begins with a software process which processes an STL file (stereolithography file format), mathematically slicing and orienting the model for the build process. If required, support structures may be generated. The machine may dispense multiple materials to achieve different goals: For example, one may use one material to build up the model and use another as a soluble support structure,[3] or one could use multiple colors of the same type of thermoplastic on the same model.

The model or part is produced by extruding small flattened strings of molten material to form layers as the material hardens immediately after extrusion from the nozzle

A plastic filament or metal wire is unwound from a coil and supplies material to an extrusion nozzle which can turn the flow on and off. There is typically a worm-drive that pushes the filament into the nozzle at a controlled rate.

The nozzle is heated to melt the material. The thermoplastics are heated past their glass transition temperature and are then deposited by an extrusion head.

The nozzle can be moved in both horizontal and vertical directions by a numerically controlled mechanism. The nozzle follows a tool-path controlled by a computer-aided manufacturing (CAM) software package, and the part is built from the bottom up, one layer at a time. Stepper motors or servo motors are typically employed to move the extrusion head. The mechanism used is often an X-Y-Z rectilinear design, although other mechanical designs such as deltabot have been employed.

Although as a printing technology FDM is very flexible, and it is capable of dealing with small overhangs by the support from lower layers, FDM generally has some restrictions on the slope of the overhang, and cannot produce unsupported stalactites.

Myriad materials are available, such as Acrylonitrile Butadiene Styrene ABS, Polylactic acid PLA, Polycarbonate PC, Polyamide PA, Polystyrene PS, lignin, rubber, among many others, with different trade-offs between strength and temperature properties. Recently a German company demonstrated for the first time the technical possibility of processing granular PEEK into filament form and 3D printing parts from the filament material using FDM-technology.[4]

31.3 Commercial applications

FDM, a prominent form of rapid prototyping, is used for prototyping and rapid manufacturing. Rapid prototyping facilitates iterative testing, and for very short runs, rapid manufacturing can be a relatively inexpensive alternative.[5]

FDM uses the thermoplastics ABS, ABSi, polyphenylsulfone (PPSF), polycarbonate (PC), and Ultem 9085, among others. These materials are used for their heat resistance properties. Ultem 9085 also exhibits fire retardancy making it suitable for aerospace and aviation applications.

FDM is also used in prototyping scaffolds for medical tissue engineering applications.[6]

31.4 Free applications

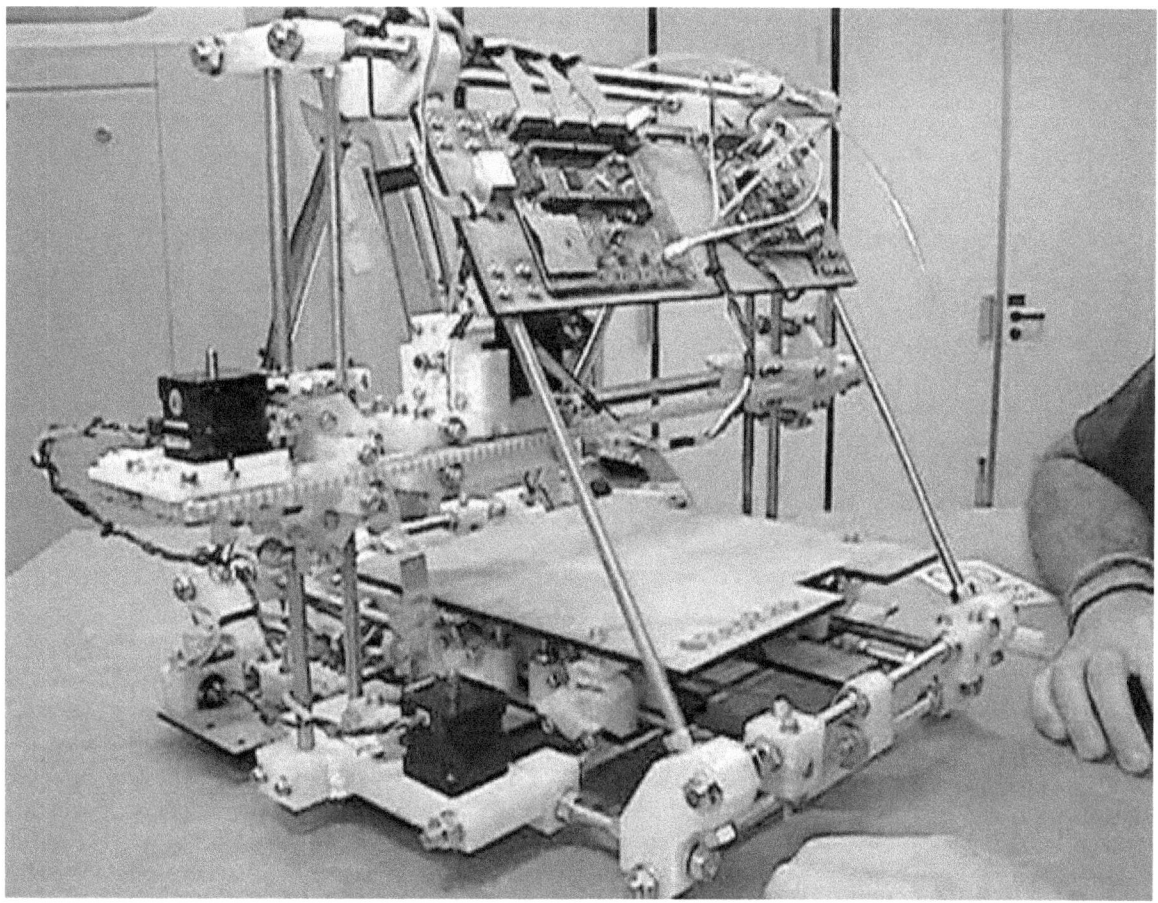

RepRap version 2.0 (Mendel)

Several projects and companies are making efforts to develop affordable 3D printers for home desktop use. Much of this work has been driven by and targeted at DIY/enthusiast/early adopter communities, with additional ties to the academic and hacker communities.[7]

RepRap is one of the longest running projects in the desktop category. The RepRap project aims to produce a free and open source hardware (FOSH) 3D printer, whose full specifications are released under the GNU General Public License, and which is capable of replicating itself by printing many of its own (plastic) parts to create more machines.[8][9] RepRaps have already been shown to be able to print circuit boards[10] and metal parts.[11][12]

Fab@Home is the other opensource hardware project for DIY 3D printers.

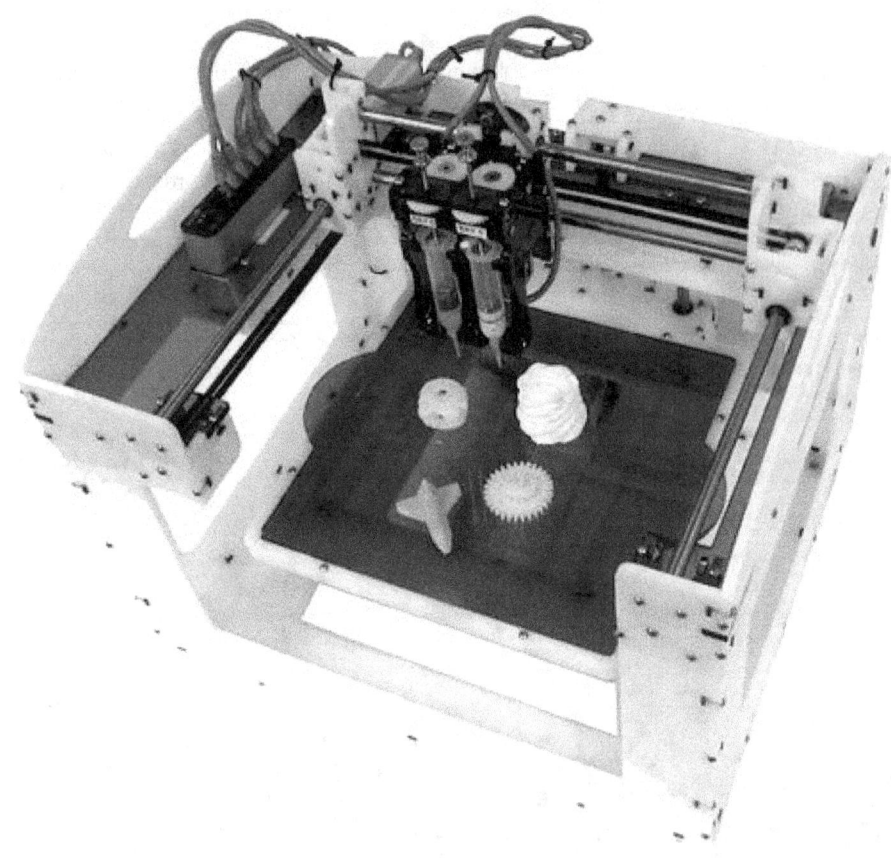

Fab@Home Model 2 (2009)

Because of the FOSH aims of RepRap, many related projects have used their design for inspiration, creating an ecosystem of related or derivative 3D printers, most of which are also open source designs. The availability of these open source designs means that variants of 3D printers are easy to invent. The quality and complexity of printer designs, however, as well as the quality of kit or finished products, varies greatly from project to project. This rapid development of open source 3D printers is gaining interest in many spheres as it enables hyper-customization and the use of public domain designs to fabricate open source appropriate technology. This technology can also assist initiatives in sustainable development since technologies are easily and economically made from resources available to local communities.[13][14]

The cost of 3D printers has decreased dramatically since about 2010, with machines that used to cost $20,000 now costing less than $1,000.[15] For instance, as of 2013, several companies and individuals are selling parts to build various RepRap designs, with prices starting at about €400 / US$500.[16] The open source Fab@Home project[17] has developed printers for general use with anything that can be squirted through a nozzle, from chocolate to silicone sealant and chemical reactants. Printers following the project's designs have been available from suppliers in kits or in pre-assembled form since 2012 at prices in the US$2000 range.

The LulzBot 3D printers manufactured by Aleph Objects are another example of an open-source application of fused deposition modeling technology. The flagship model in the LulzBot line, the TAZ printer takes inspiration for its design from the RepRap Mendel90 and Prusa i3 models. The LulzBot 3D printer is currently the only printer on the market to have received the "Respects Your Freedom" certification from the Free Software Foundation.[18]

Printing in progress in a Ultimaker 3D printer during Mozilla Maker party, Bangalore

31.5   See also

- 3D printing

- 3D printer extruder

- CEL Robox

- Direct metal laser sintering

- Fab lab

- Fab@Home

- Hyrel 3D

- MakerBot Industries

- Printrbot

- Rapid prototyping

- RepRap Project

- Selective laser sintering

- Stereolithography

- Von Neumann universal constructor

- Robo 3D

- G-code

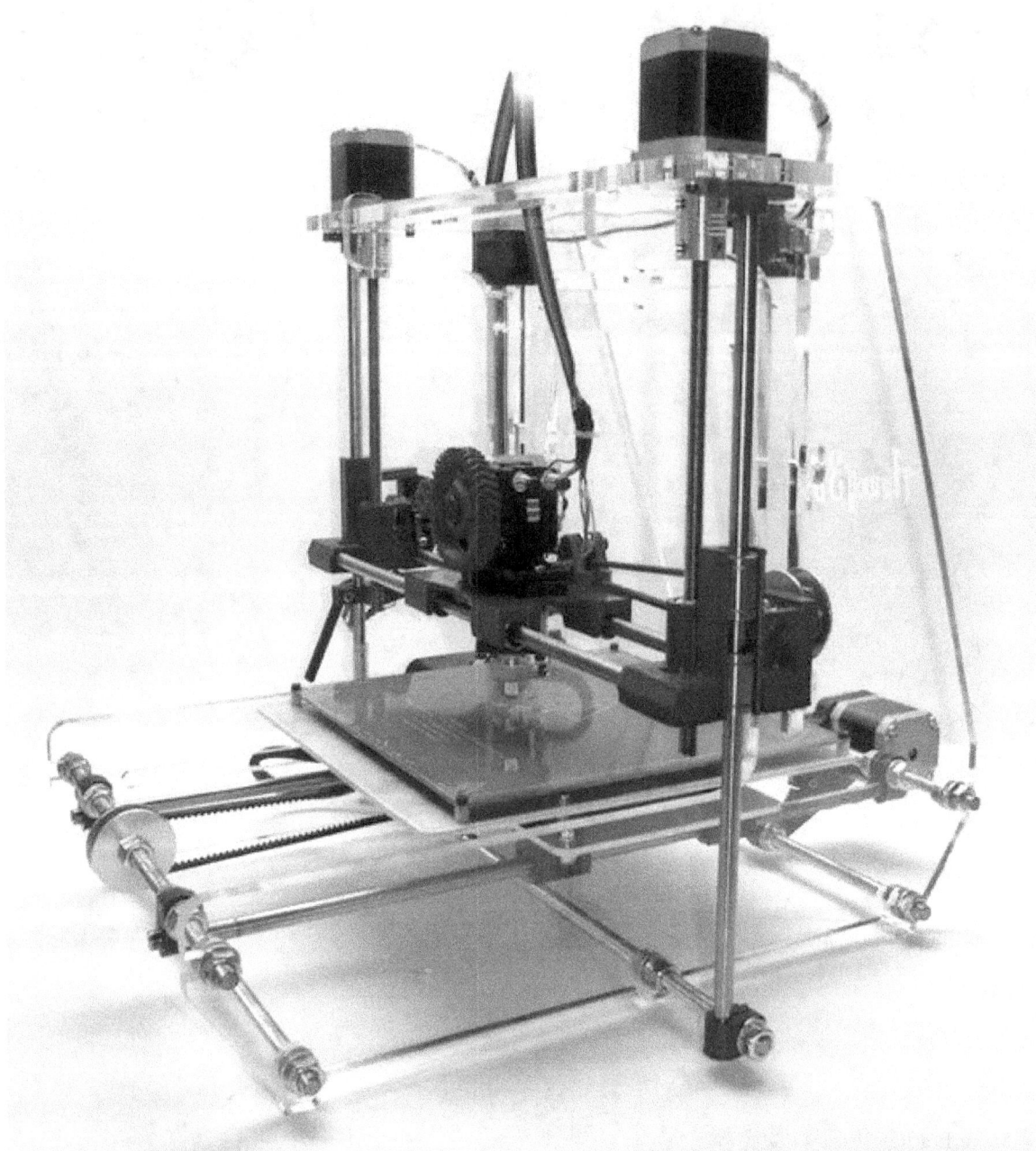

Airwolf 3D AW3D v.4 (Prusa)

31.6 References

[1] http://rpworld.net/cms/index.php/additive-manufacturing/rp-rapid-prototyping/fdm-fused-deposition-modeling-.html

[2] Chee Kai Chua; Kah Fai Leong, Chu Sing Lim (2003). *Rapid Prototyping*. World Scientific. p. 124. ISBN 9789812381170.

[3] http://www.engr.mun.ca/~{}kmay/CleanStation/MSDSP400SCWaterWorks_US.pdf

[4] "3dprint.com, PEEK being 3D-printed". *3dprint.com*. March 21, 2015. Retrieved March 26, 2015.

[5] https://books.google.com/books?id=GUhhs3MnQR4C

[6] Ferry Melchels et al 2011 Biofabrication 3 034114 doi:10.1088/1758-5082/3/3/034114

[7] Kalish, Jon. "A Space For DIY People To Do Their Business (NPR.org, November 28, 2010)". Retrieved 2012-01-31.

[8] Jones, R., Haufe, P., Sells, E., Iravani, P., Olliver, V., Palmer, C., & Bowyer, A. (2011). Reprap-- the replicating rapid prototyper. Robotica, 29(1), 177-191.

[9] "Open source 3D printer copies itself". Computerworld New Zealand. 2008-04-07. Retrieved 2013-10-30.

[10] RepRap blog 2009 visited 2/26/2014

[11] An Inexpensive Way to Print Out Metal Parts - The New York Times

[12] Gerald C. Anzalone, Chenlong Zhang, Bas Wijnen, Paul G. Sanders and Joshua M. Pearce, " Low-Cost Open-Source 3-D Metal Printing" *IEEE Access*, 1, pp.803-810, (2013). doi: 10.1109/ACCESS.2013.2293018

[13] Pearce, Joshua M.; et al. "3-D Printing of Open Source Appropriate Technologies for Self-Directed Sustainable Development (Journal of Sustainable Development, Vol.3, No. 4, 2010, pp. 17–29)". Retrieved 2012-01-31.

[14] Tech for Trade, 3D4D Challenge

[15] Disruptions: 3-D Printing Is on the Fast Track – NYTimes.com

[16] www.3ders.org. "3D printers list with prices". 3ders.org. Retrieved 2013-10-30.

[17] New Scientist magazine: Desktop fabricator may kick-start home revolution, 9 January 2007

[18] Gay, Joshua (29 Apr 2013). "Aleph Objects". *fsf.org*. Free Software Foundation, Inc. Retrieved 2 April 2015.

31.7　Further reading

- "Results of Make Magazine's 2015 3D Printer Shootout". docs.google.com. Retrieved 1 June 2015.

- "Evaluation Protocol for Make Magazine's 2015 3D Printer Shootout". makezine.com. Retrieved 1 June 2015.

- Stephens, Brent; Parham Azimia; Zeineb El Orcha; Tiffanie Ramos (November 2013). "Ultrafine Particle Emissions from Desktop 3D Printers". *Atmospheric Environment* **79**: 334–339. doi:10.1016/j.atmosenv.2013.06.050. Retrieved 13 August 2013.

- "How Fused Deposition Modeling Works". THRE3D.com. Retrieved 7 February 2014.

- "3D Printing process and How FDM technology works Video". homeshop3dprinting.com. Retrieved 4 June 2014.

- "Complete list of G-code used by 3D printer's firmware of RepRap project". RepRap.org. Retrieved 26 August 2015.

Chapter 32

Fused filament fabrication

Fused filament fabrication is a 3D printing process that uses a continuous filament of a thermoplastic material. This is fed from a large coil, through a moving, heated **printer extruder head**. Molten material is forced out of the print head's nozzle and is deposited on the growing workpiece. The head is moved, under computer control, to define the printed shape. Usually the head moves in layers, moving in two dimensions to deposit one horizontal plane at a time, before moving slightly upwards to begin a new slice. The speed of the extruder head may also be controlled, to stop and start deposition and form an interrupted plane without stringing or dribbling between sections.

Fused filament printing is now the most popular process (by number of machines) for hobbyist-grade 3D printing. As other techniques, such a photopolymerisation and powder sintering, may offer better results at greater cost, they still dominate commercial printing.

The **3D printer head** or **3D printer extruder** is a part in material extrusion-type printing responsible for raw material melting and forming it into a continuous profile. A wide variety of materials are extruded, including thermoplastics such as *acrylonitrile butadiene styrene* (ABS), *polylactic acid* (PLA), *high-impact polystyrene* (HIPS), *thermoplastic polyurethane* (TPU), *aliphatic polyamides* (nylon),[1] and recently also PEEK. [2] You can also extrude paste-like materials including ceramic and chocolate.[3]

32.1   Introduction

Additive manufacturing (AM), also referred to as 3D printing, involves manufacturing a part by depositing material layer by-layer. There is a wide array of different AM technologies that can make a part layer-by-layer including material extrusion, binder jetting, material jetting and directed energy deposition.[4]

These process have varied types of extuders and extrude different materials to achieve the final product using layer by layer addition of material approach. The **3D Printer Liquefier** is the component predominantly used in Material extrusion type printing.

32.1.1   Extrusion

Extrusion in 3-D printing using material extrusion involves a cold end and a hot end.

The **cold end** is part of an extruder system that pulls and feed the material from the spool, and pushes it towards the hot end. The cold end is mostly gear- or roller-based supplying torque to the material and controlling the feed rate by means of stepper motor. Thus, controlling the process rate in result.

The **hot end** is the active part which also hosts the liquefier of the 3D printer that melts the filament. It allows the molten plastic to exit from the small nozzle to form a thin and tacky bead of plastic that will adhere to the material it is laid on. Hot end consists of heating chamber and nozzle.T he hole in the tip (nozzle) has a diameter of between 0.3 mm and 1.0 mm. Different types of nozzles and heating methods are used depending upon the material to be printed.[5]

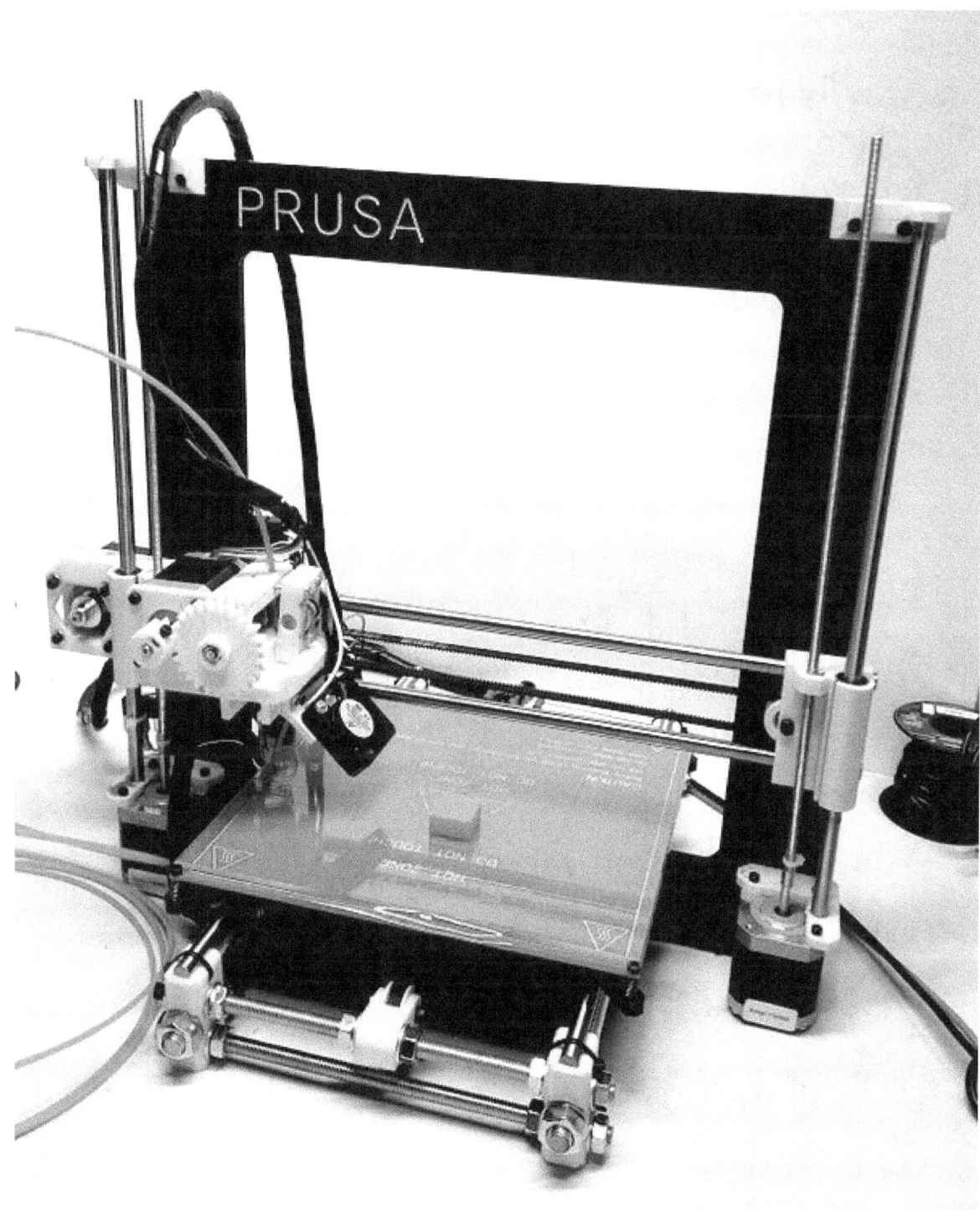

Prusa, a simple fused filament printer

Some type of 3-D printing machines can have a different type of extrusion system which may not have heating chamber and the heat is supplied from other source like laser.

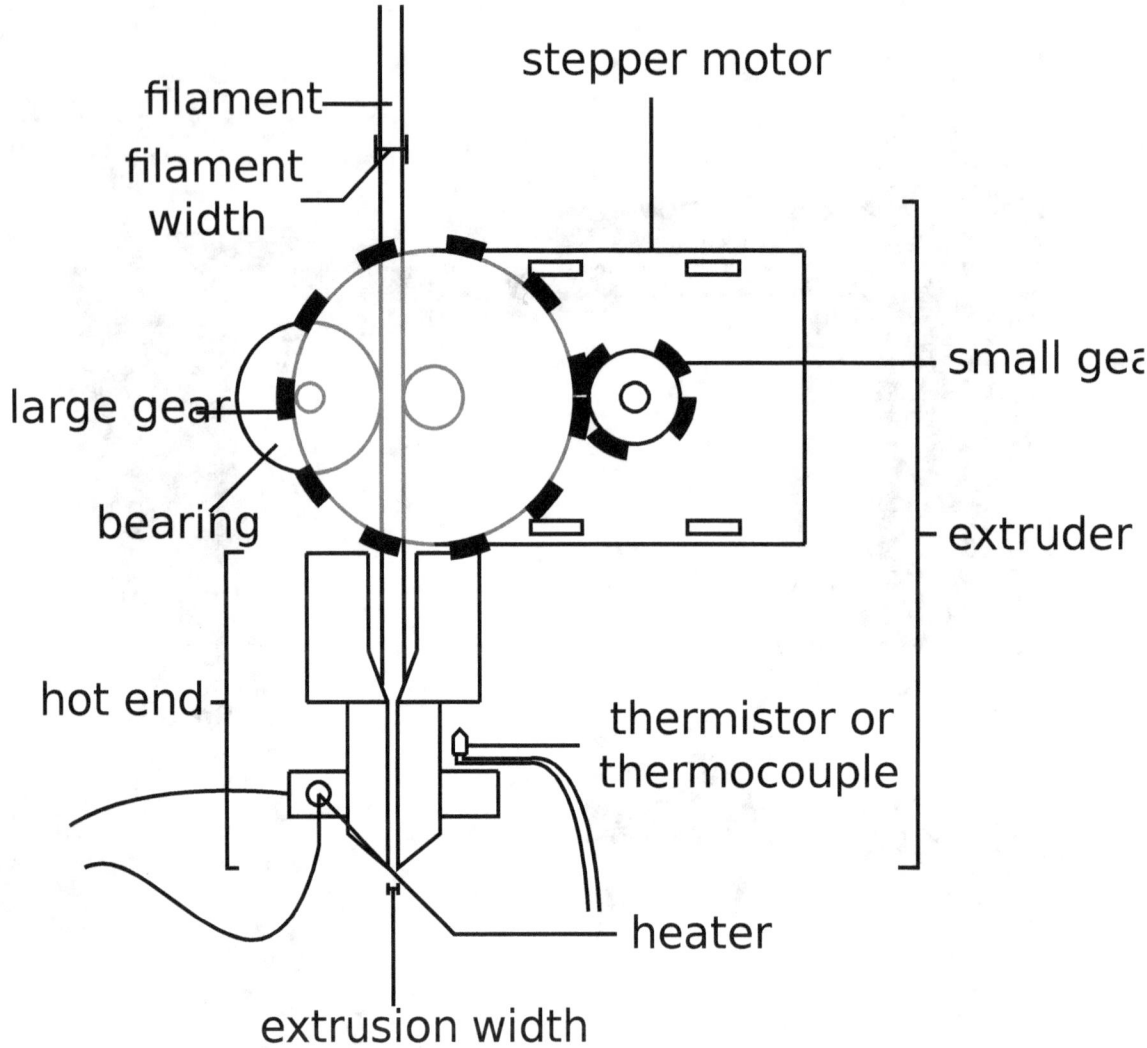

Illustration of an extruder, that shows how all parts are named.

32.1.2 Extruder mount to rest of machine

The ways extruders are mounted on the rest of the machine have evolved over time into informal mounting standards. These informal standards include:[5]

- the Vertical X Axis Standard,

- the Quick-fit extruder mount,

- the OpenX mount,

etc.

Such de-facto standards allows new extruder designs to be tested on existing printer frames, and new printer frame designs to use existing extruders.

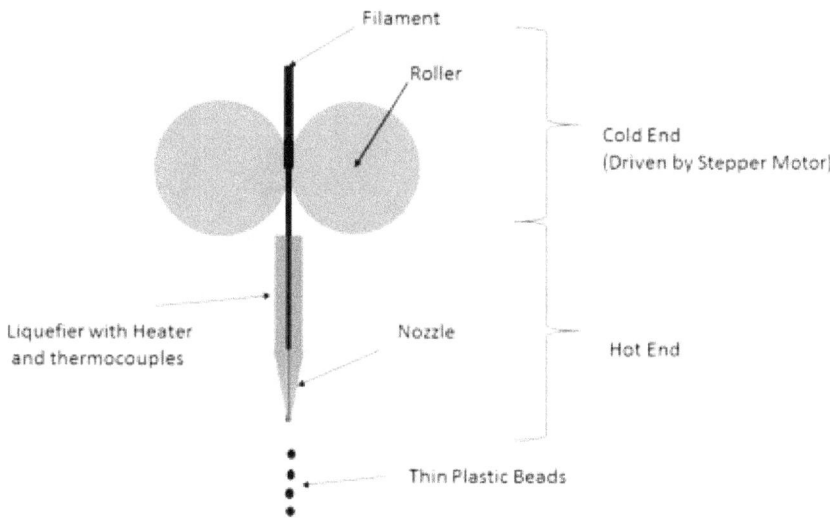

3-D Printer Extruder

32.2 Process

Flow geometry of the extruder, heating method and the melt flow behavior of a non-Newtonian fluid are of main consideration in the part.

A plastic filament is supplied from a reel commercially available or home made and fed into a heated liquefier where it is melted. This melt is then extruded by a nozzle while the incoming filament, still in solid phase, acts as a "**plunger**."

The nozzle is mounted to a mechanical stage, which can be moved in the **xy** plane. As the nozzle is moved over the table in a prescribed geometry, it deposits a thin bead of extruded plastic, called "**roads**" which solidify quickly upon contact with substrate and/or roads deposited earlier.[6]

Solid layers are generated by following a rasterizing motion where the roads are deposited side by side within an enveloping domain boundary.

Once a layer is completed, the platform is lowered in the **z direction** in order to start the next layer. This process continues until the fabrication of the object is completed.

For Successful bonding of the roads in the process control of the thermal environment is necessary. Therefore, the system is kept inside a chamber, maintained at a temperature just below the melting point of the material being deposited.

32.3 Physics

During extrusion the thermoplastic filament is introduced by mechanical pressure from Rollers, into the liquefier, where it melts and is then extruded. The rollers are the only drive mechanism in the material delivery system, therefore filament is under tensile stress upstream to the roller and under compression at the downstream side acting as a plunger. Therefore compressive stress is the driving force behind the extrusion process.

The force required to extrude the melt must be sufficient to overcome the pressure drop across the system, which strictly depends on the viscous properties of the melted material and the flow geometry of the liquefier and nozzle. The melted material is subjected to shear deformation during the flow. Shear thinning behavior is observed in most of the materials used in this type of 3-D printing. This is modeled using power law for generalized Newtonian fluids.

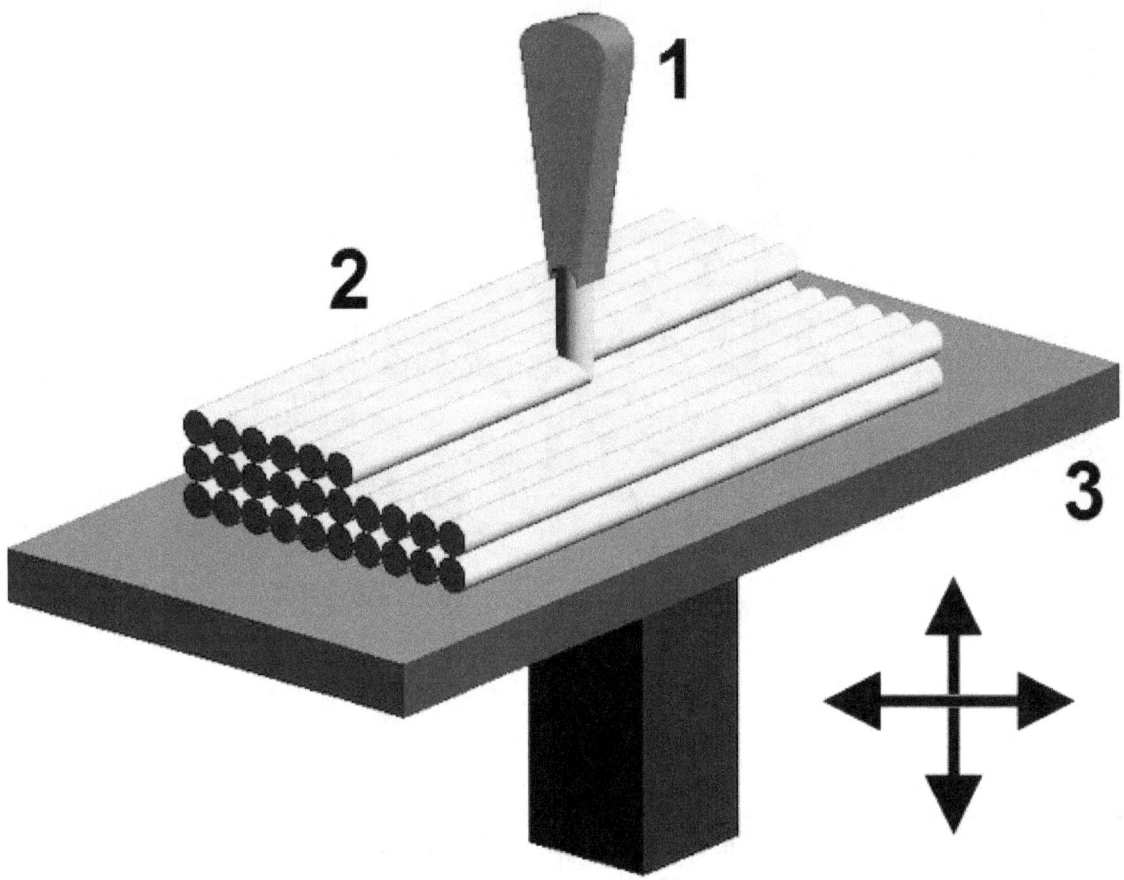

Process: 1 – 3D Printer Extruder, 2 – deposited material (modeled part), 3 – controlled movable table

The temperature is regulated by heat input from electrical coil heaters. The system continuously adjusts the power supplied to the coils according to the temperature difference between the desired value and the value detected by the thermocouple. Forming a Negative Feedback loop. This is similar to Heat Flow rate in Cylindrical Pipe.

32.4 Types and uses

[7]

32.5 Development

Customer-driven product customization and demand for cost and time savings has increased interest in agility of manufacturing process. This has led to improvements in RP technologies and in particularly of Fused Deposition Modeling.[6] The Development of Extruders is going rapidly because of open source 3-D printer movement caused by products like RepRap. Consistent improvements are seen in the from increased heating temperature of liquefier, the over-all control and precision of the process and improved support for wide variety of materials to print, including ceramics.

The ways extruders are mounted on the machine has also evolved over time into informal mounting standards. These informal standards include the Vertical X Axis Standard, the Quick-fit extruder mount, the OpenX mount, etc.

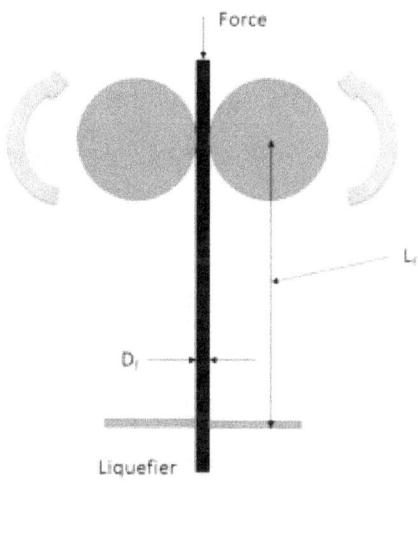

Driving Force Due
to Rollers

3D Printer Extruder Driving Force. Where D_f is Diameter of Filament and L_f is Length of filament

32.6   Print head kinematics

The majority of fused filament printers follow the same basic design. A flat bed is used as the starting point for the print workpiece. A gantry above this carries the moving print head. The gantry design is optimised for movement mostly in the horizontal X & Y directions, with a slow climb in the Z direction as the piece is printed. Stepper motors drive the movement through either leadscrews or toothed belt drives. It is common, owing to the differences in movement speed, to use toothed belts for the X,Y drives and a leadscrew for Z. Some machines also have X axis movement on the gantry, but move the bed (and print job) for Y. As, unlike laser cutters, head movement speeds are low, stepper motors are universally used and there is no need to use servomotors instead.

Many printers, originally those influenced by the RepRap project, make extensive use of 3D printed components in their own construction. These are typically printed connector blocks with a variety of angled holes, joined by cheap steel threaded rod. This makes a construction that is cheap and easy to assemble, easily allows non-perpendicular framing joints, but does require access to a 3D printer. The notion of 'bootstrapping' 3D printers like this has been something of a dogmatic theme within the RepRap designs. The lack of stiffness in the rod also requires either triangulation, or gives the risk of a gantry structure that flexes and vibrates in service, reducing print quality.

Many machines now use box-like semi-enclosed frames of either laser-cut plywood, plastic or pressed steel sheet. These are cheap, rigid and can also be used as the basis for an enclosed print volume, allowing temperature control within it to control warping of the print job.

A handful of machines use polar coordinates instead, usually machines optimised to print objects with circular symmetry. These have a radial gantry movement and a rotating bed. Although there are some potential mechanical advantages to this design for printing hollow cylinders, their different geometry and the resulting non-mainstream approach to print planning still keeps them from being popular as yet. Although it is an easy task for a robot's motion planning to convert from Cartesian to polar coordinates, gaining any advantage from this design also requires the print slicing algorithms to be aware of the rotational symmetry from the outset.

RepRap-type printer

32.6.1 Rostock printers

A different approach is taken with 'Rostock' pattern printers, based on a delta robot mechanism.[8] These have a large open print volume with a three-armed delta robot mounted at the top. This design of robot is noted for its low inertia and ability for fast movement over a large volume. Stability and freedom from vibration when moving a heavy print head on the end of spindly arms is a technical challenge though. This design has mostly been favoured as a means of gaining a large print volume without a large and heavy gantry.

As the print head moves the distance of its filament from storage coil to head also changes. This tugging on the filament is another technical challenge to overcome, if it is not to affect print quality.

Printing by a large delta robot printer

32.7   See also

- Ball bearing
- Fused deposition modeling
- Methacrylate
- Plastics extrusion
- Rod
- RAMPS
- Stepper motor
- Spindle
- Thermistor
- Thermocouple

32.8   References

[1] "RepRap Wiki Category:Thermoplastics". Retrieved 2 November 2014.

[2] "3dprint.com, PEEK being 3D-printed". *3dprint.com*. March 21, 2015. Retrieved March 26, 2015.

[3] "Universal Paste extruder – Ceramic, Food and Real Chocolate 3D Printing". Retrieved 2 November 2014.

[4] Brett P. Conner∗, Guha P. Manogharan, Ashley N. Martof, Lauren M. Rodomsky, Caitlyn M. Rodomsky, Dakesha C. Jordan, James W. Limperos; Manogharan; Martof; Rodomsky; Rodomsky; Jordan; Limperos (2014). "Making sense of 3-D printing: Creating a map of additive manufacturing products and services". *Addit Manuf*. doi:10.1016/j.addma.2014.08.005.

[5] "FDM Extruders". Retrieved 24 October 2014.Reprap extruders

[6] Selc¸uk Gu¨c¸eri, Maurizio Bertoldi; GüçEri; Bertoldi (2014). "Liquefier Dynamics in Fused Deposition". *Journal of Manufacturing Science and Engineering* **126** (2): 237. doi:10.1115/1.1688377.Liquefier Dynamics in Fused Deposition

[7] "Hot End Reprap". Retrieved 24 October 2014.

[8] "Rostock". *RepRap*.

Chapter 33

IMakr

iMakr Store in London

iMakr is an independent value added reseller of 3D printers and 3D printed related goods. iMakr currently operates two physical dedicated 3D printing stores, one in London and one New York[1] as well as an extensive online and phone sales service that delivers worldwide.

33.1 History

33.1.1 Founding

iMakr was established in April 2013 with the opening of its first store on Clerkenwell Road, London.[2]

33.1.2 Expansion

Following the success of its first London store, the company opened a second flagship store in Manhattan in June 2014.[3]

33.1.3 MyMiniFactory

In June 2013, iMakr launched MyMiniFactory,[4] a website dedicated to the free sharing of 3D printable files.

33.2 Services

In addition to selling best-in-class 3D printers and 3D printing related products, **iMakr** operates a 3D scanning service and prints full colour figurines of their customers.[5][6] They have also used 3D scanning to offer their customers 3D printed candles, in partnership with French department store le Bon Marché.[7] and chocolates in partnership with Rococo Chocolates.[8] Other services provided by iMakr include print on demand, consulting and weekly free 3D printing training sessions. iMakr also participates regularly in prominent events and exhibitions where it demonstrates 3D Printing in an effort to help people understand what this technology can do for them.

33.3 References

[1] Store Locator

[2] *http://makerfaireelephantandcastle.com/* The world's largest 3D printing store…

[3] *http://www.inside3dp.com* iMakr Opens NY Store Following London Success

[4] *http://solidsmack.com/* iMakr Opening My Mini Factory to Provide Free 3D Printable Files (and Sell Yours.)

[5] *The Guardian.com* Could the mini-me make 3D-printing mainstream?

[6] *http://mini-you.co.uk/* iMakr's 3D scan service: Mini-You

[7] *http://microfabricator.com* Your Scanned Head + Candle = Scandle, a 3D Printed Wax Mini You

[8] *http://www.3ders.org* iMakr & Rococo offer 3D chocolate selfie at Harvey Nichols London

33.4 External links

- iMakr official e-commerce site

- MyminiFactory: iMakr's 3D printable files sharing website

- MyminiFactory: iMakr's Mini-You website

Chapter 34

Laminated object manufacturing

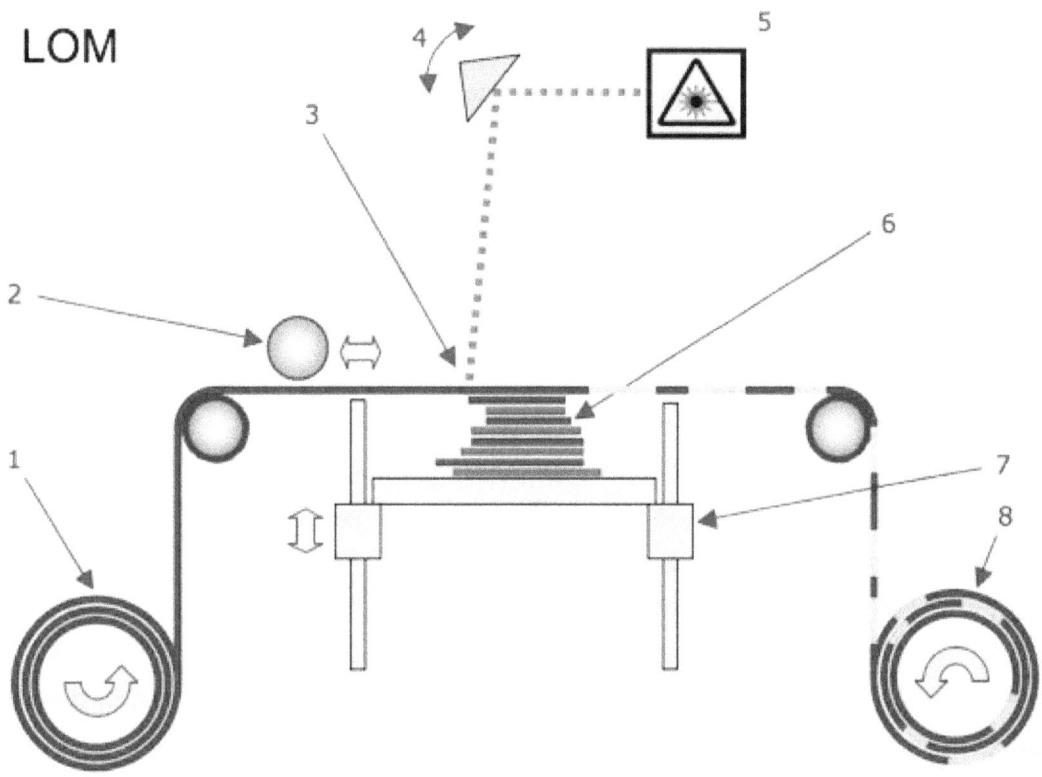

LOM

Laminated object manufacturing: 1 Foil supply. 2 Heated roller. 3 Laser beam. 4. Scanning prism. 5 Laser unit. 6 Layers. 7 Moving platform. 8 Waste.

Laminated object manufacturing (**LOM**) is a rapid prototyping system developed by Helisys Inc. (Cubic Technologies is now the successor organization of Helisys) In it, layers of adhesive-coated paper, plastic, or metal laminates are successively glued together and cut to shape with a knife or laser cutter. Objects printed with this technique may be additionally modified by machining or drilling after printing. Typical layer resolution for this process is defined by the material feedstock and usually ranges in thickness from one to a few sheets of copy paper.[1]

The process is performed as follows:

1. Sheet is adhered to a substrate with a heated roller.

2. Laser traces desired dimensions of prototype.

3. Laser cross hatches non-part area to facilitate waste removal.

4. Platform with completed layer moves down out of the way.

5. Fresh sheet of material is rolled into position.

6. Platform downs into new position to receive next layer.

7. The process is repeated.

Note:

- Low cost due to readily available raw material

- Paper models have wood like characteristics, and may be worked and finished accordingly

- Dimensional accuracy is slightly less than that of stereolithography and selective laser sintering but no milling step is necessary.

- Relatively large parts may be made, because no chemical reaction is necessary.[2][3]

34.1 References

[1] "How Laminated Object Manufacturing Works". THRE3D.com. Retrieved 3 February 2014.

[2] "Laminated Object Manufacturing." April 10, 2006.https://web.archive.org/web/20100102182152/http://home.att.net/~{}castleisland/lom.htm (accessed April 19, 2008).

[3] "Rapid Prototyping: LOM." http://www.efunda.com/processes/rapid_prototyping/lom.cfm (accessed June 8, 2012).

Chapter 35

Laser engineered net shaping

Laser engineered net shaping or **LENS** (Also known by the non-proprietary name "**Laser Powder Forming**") is an additive manufacturing technology developed for fabricating metal parts directly from a computer-aided design (CAD) solid model by using a metal powder injected into a molten pool created by a focused, high-powered laser beam. This technique is also equivalent to several trademarked techniques that have the monikers Direct Metal Deposition (DMD), and Laser consolidation (LC). Compared to processes that use powder beds, such as Selective Laser Melting (SLM), objects created with this technology can be substantially larger, even up to several feet long.[1]

A high power laser is used to melt metal powder supplied coaxially to the focus of the laser beam through a deposition head. The laser beam typically travels through the center of the head and is focused to a small spot by one or more lenses. The X-Y table is moved in raster fashion to fabricate each layer of the object. The head is moved up vertically as each layer is completed. Metal powders are delivered and distributed around the circumference of the head either by gravity, or by using a pressurized carrier gas. An inert shroud gas is often used to shield the melt pool from atmospheric oxygen for better control of properties, and to promote layer to layer adhesion by providing better surface wetting.

This process is similar to other 3D fabrication technologies in its approach in that it forms a solid component by the layer additive method. The LENS process can go from metal and metal oxide powder to metal parts, in many cases without any secondary operations. LENS is similar to selective laser sintering, but the metal powder is applied only where material is being added to the part at that moment. It can produce parts in a wide range of alloys, including titanium, stainless steel, aluminum, and other specialty materials; as well as composite and functionally graded materials. Primary applications for LENS technology include repair & overhaul, rapid prototyping, rapid manufacturing, and limited-run manufacturing for aerospace, defense, and medical markets. Microscopy studies show the LENS parts to be fully dense with no compositional degradation. Mechanical testing reveals outstanding as-fabricated mechanical properties.

The process can also make "near" net shape parts when it's not possible to make an item to exact specifications. In these cases post production light machining, surface finishing, or heat treatment may be applied to achieve end compliance.

35.1 External links

- Sandia National Laboratory LENS article.

35.2 References

[1] "How Laser Powder Forming Works". THRE3D.com. Retrieved 11 February 2014.

Chapter 36

Laser sintering of gold

Design by Towe Norlén

Laser sintering of gold is a jewellery manufacturing technique[1] first developed by Towe Norlén and Lena Thorsson.[2]

Laser sintering [3] of gold starts with gold powder, fine as flour. A laser beam sinters (melts) the gold flour locally in an extremely small point, and any shape may be 'drawn' precisely with the laser beam, in three dimensions. When the gold object is finished, it is gently brushed from the leftover gold flour, in much the same way as in an archaeological dig.

The result is a gold object of virtually any shape, and with higher quality (greater surface density) gold,[4] than that possible to achieve with casting. Moreover, laser sintering circumvents the weakening and surface-deforming mounting process, because the item of jewellery is manufactured in a single piece. Also, jewellery design may be expanded and individualised, as in principle any shape is possible, which facilitates uniqueness and personalized design.

36.1 References

[1] 1. Wojtkielo Snyder, T (2013) Exciting times. MJSA journal 3/2013:1

[2] http://technical-journal.thegoldsmiths.co.uk/wp-content/uploads/2013/08/Tech-Bulletin-No3.pdf

[3] 2. Wojtkielo Snyder, T (2013) Imagine that: additive manufacturing offers new opportunities, MJSA journal 3/2013:22-30

[4] Wojtkielo Snyder, T (2012) Growing up: additive manufacturing beams down to the jewelry scene. MJSA journal 11/2012:18-21

36.2 External links

Websites

- http://www.rapidnews.com/TCT-presentations-2012/TCT%20Cookson%20Precious%20Metals.pdf

- http://www.housecreative.co.uk/casestudy_cooksondmls.php

- http://media.messe.ch/epaper/BASELWORLD/2013/machines_and_supply_industry/page46.html#/44

- http://www.cooksongold-emanufacturing.com/

- http://www.metal-powder.net/view/29659/eos-unveils-new-laser-sintering-machine/

- http://www.detekt.com.tw/download/eos/7M⬜⬜⬜⬜⬜⬜/EOSINT%20M%20Materials%20for%20Direct%20Metal%20Laser-Sintering%20(DMLS).pdf

Videos:

- Video on YouTube

- Video on YouTube

- Video on YouTube

- Video on YouTube

Chapter 37

List of 3D printer manufacturers

Below is a list of 3D Printer manufacturers listed by company name and location. 3D printers are a type of industrial robot that is able to print 3D models using successive layers of material.

37.1 0–9

- 3DISON – Wieringerwerf, Netherlands
- 3D Stuffmaker – Miami, Florida, USA
- 3D Systems – Rock Hill, South Carolina, USA *Note: Brand name is Cubify
- 3DGence – Poland

37.2 A

- AIO Robotics – Los Angeles, California, USA
- Arcam AB – Mölndal, Sweden
- Aurora Technology - Shenzhen, China
- AutoDesk – San Rafael, California, USA
- Airwolf 3D – Costa Mesa, California, USA
- Aleph Objects – Loveland, Colorado, USA *Note: Brand name is Lulzbot

37.3 B

- Bee Very Creative - Gafanha d'Aquém, Portugal
- Beijing Tiertime Technology Co – Beijing, China *Note: USA brand is Afina, Europe brand is Up
- be3D – Czechoslovakia
- Bob's CNC – Independence, Kansas, USA
- B9Creations – Rapid City, South Dakota, USA
- BigRep – Berlin, Germany

37.4   C-F

- Carima – Seoul, Korea

- Cel – Portishead, United Kingdom

- Deezmaker - Pasadena, California, USA

- deltamaker - Orlando, Florida, USA

- dfrobot - Shanghai, China

- envisionTEC – Gladbeck, Germany

- EOS – Krailling, Germany

- FELIXprinters – IJsselstein, Netherlands

- FlashForge – Jinhua, China *Note: OEM for Dremmel

- Formlabs – Somerville, Massachusetts, USA

- Fusion3 – Greensboro, North Carolina, USA

37.5   G-L

- gCreate – New York City, New York, USA

- HIC Technology – Shenzhen, China *Brand name is HICTOP

- Hyrel 3D – Norcross, Georgia, USA

- iBox Printers - Melbourne, Florida, USA

- Ira3D – Italy

- Kudo3D – Pleasanton, California, USA

- Leapfrog 3D Printers – Alphen aan den Rijn, Netherlands

37.6   M

- M3D – Fulton, Maryland, USA

- MAKEX - Zhejiang, China

- MaherSoft – Mumbai, India

- Makeblock - Shenzhen, China

- MakerBot Industries – New York City, New York, USA

- MarkForged – Waltham, Massachusetts, USA

- MakerGear – Beachwood, Ohio, USA

- Materialise NV – Leuven, Belgium

- MBot – China

- Mcor Technologies – Dunleer, Ireland

- MingDa – Shenzhen, China

- Moment – Seoul, Korea

- Mundo Reader – Las Rozas, Spain *Brand is BQ Witbox

37.7   N-Q

- NW RepRap – Arizona, USA

- Pirate3D – Singapore

- Polar 3D – Cincinnati, Ohio, USA

- Printrbot – Lincoln, California, USA

- Print - Rite - Hong Kong *Brand is CoLido

- Q3D – China

- Qualup SAS – France

37.8   R

- Raise3D – China

- RepRapPro – Bristol, United Kingdom

- Robo3D – San Diego, California, USA

- Ruian Qidi Technology Co. – Ruian, Zhejiang, China

37.9   S

- Sciaky, Inc. – Chicago, Illinois, USA

- SeeMeCNC – Goshen, Indiana, USA

- Shaoxing Bibo Automatic Equipment - Shaoxing, China

- Shenzhen Weistek Co. – Shenzhen, China

- Solido 3D – Italy

- Solidoodle – New York City, New York, USA

- Something 3D – Israel

- Stacker 3D - Plymouth, Minnesota, USA

- Stratasys – Minneapolis, Minnesota, USA

- Sunruy Technology – China

37.10   T

- Tiko – Niagara Falls, New York, USA

- Tri-Tech 3D – Staffordshire, United Kingdom

- Tinkerine – Vancouver, British Columbia, Canada

- TripodMaker – Brussels, Belgium

- Type A Machines – San Leandro, California, USA

37.11   U-Z

- Ultimaker – Geldermalsen, Netherlands

- Velleman – Belgium

- voxel8 – Somerville, Massachusetts, USA

- Voxeljet – Friedberg, Germany

- Wanhao – Zhejiang, China

- Wiiboox – China

- xyzPrinting – Taiwan

- ZeePro - San Francisco, California, USA

- Zmorph – Wrocław, Poland

- Zortrax – Olsztyn, Poland

- Zhuhai CTC Electronic - Zhuhai City, Guangdong, China

Chapter 38

Local Motors

Local Motors Rally Fighter

Local Motors is an American motor vehicle manufacturing company focused on low-volume manufacturing of open-source motor vehicle designs using multiple microfactories. They were founded in 2007 with headquarters in Phoenix, Arizona.[1] Current products range from the Rally Fighter automobile and Racer motorcycle; through various electric bicycles, tricycles, and children's ride-on toy cars; to radio-controlled model cars and skateboards. They 3D print some components to the vehicles and toys they sell.[2] Rally Fighters have used co- creation techniques to its designing phase.It is a technique that is used by the companies to enhance in new product development.

157

38.1   Community

Local Motors web site is an important community about engine vehicles innovation. The content is created by the users of the community who discuss about design, engineering and building of innovative engine vehicles. Members contribute with their own ideas and projects which are discussed with the community. Due to the co-design of the vehicles with the contribution of their worldwide members the firm has reduced the time and cost of the development of their engine vehicles.

38.2   Strati

Main article: Strati (automobile)

Local Motors in collaboration with Cincinnati Incorporated and Oak Ridge National Laboratory manufactured Strati, the world's first 3D printed electric car.[3] The printer was set up by Cincinnati Incorporated with the printing that took 44 hours to complete with live audience at the 2014 International Manufacturing Technology Show in McCormick Place, Chicago.[4] They are hoping for production by 2015.

38.3   See also

- Open design

38.4   References

[1] "About Local Motors", Local Motors Website

[2] "Vehicles", Local Motors Website

[3] Gastelu, Gary (3 July 2014). "Local Motors 3D-printed car could lead an American manufacturing revolution". Fox News. Retrieved 22 September 2014.

[4] Franklin, Dallas (15 September 2014). "Made in Chicago: World's First 3D Printed Electric Car". KFOR-TV. Retrieved 22 September 2014.

38.5   External links

- Local Motors Website

Chapter 39

Lyman filament extruder

The **Lyman filament extruder** is a device for making 3-D printer filament suitable for use in 3-D printers like the RepRap. It is named after its developer Hugh Lyman and was the winner of the Desktop Factory Competition.[1]

It is open source hardware and all of the plans can be downloaded from Thingiverse[2]

The goal of the Desktop Factory Competition[3] was to build an open source filament extruder for less than $250 in components can take ABS or PLA resin pellets, mix them with colorant, and extrude enough 1.75 mm diameter ± 0.05 mm filament that can be wrapped on a 1 kg spool. The machine must use the Attribution-ShareAlike 3.0 Unported (CC BY-SA 3.0).

The use of DIY filament extruders like the Lyman can significantly reduce the cost of printing with 3-D printers.[4] The Lyman filament extruder was designed to handle pellets, but can also be used to make filament from other sources of plastic such as post-consumer waste like other RecycleBots. Producing plastic filament from recycled plastic has a significant positive environmental impact.[5]

39.1   References

[1] Harry McCracken (March 4, 2013). "How an 83-Year-Old Inventor Beat the High Cost of 3D Printing". Time.

[2] http://www.thingiverse.com/thing:34653

[3] http://desktopfactory2012.istart.org/

[4] Christian Baechler, Matthew DeVuono, and Joshua M. Pearce, "Distributed Recycling of Waste Polymer into RepRap Feedstock" *Rapid Prototyping Journal,* **19**(2), pp. 118-125 (2013). open access

[5] M. Kreiger, G. C. Anzalone, M. L. Mulder, A. Glover and J. M Pearce (2013). Distributed Recycling of Post-Consumer Plastic Waste in Rural Areas. MRS Online Proceedings Library, 1492, mrsf12-1492-g04-06 doi:10.1557/opl.2013.258. open accesslife-cycle analysis

Chapter 40

Made In Space, Inc.

Made In Space, Inc. is an American-based company, specializing in the engineering and manufacturing of three-dimensional printers for use in microgravity. Headquartered in Mountain View, California on Moffett Field, Made In Space's 3D printer (Zero-G Printer) was the first manufacturing device in space.

40.1 History

Made In Space was founded in August 2010, by Aaron Kemmer, Jason Dunn, Mike Chen, and Michael Snyder, during that year's Singularity University Graduate Studies Program. Their primary mission is to enable humanity to become a multi-planetary species. In the spring of 2011, Made In Space created their 3D Printing Lab, at the NASA Ames Research Center, on Moffett Field, Mountain View, California. That summer, they were awarded sub-orbital flight, through NASA's Flight Opportunities Program. From July through September 2011, the Made In Space team performed over 400 microgravity test parabolas, on NASA's reduced gravity aircraft (the "Vomit Comet"), proving their 3D printing in microgravity. With this proven concept, Made In Space was award a Phase 1 SBIR, with NASA, for the design of a 3D printer to be tested on the International Space Station (ISS).[1]

In January 2013, Made In Space was awarded Phase 2 of the SBIR, by NASA, to build and flight qualify an additive manufacturing facility, with their 3D printer, for the International Space Station (ISS).[2] Phase 3 was award in February 2013, as a sole source contract to fly their 3D printer to the ISS, In May, NASA and Made In Space announced the 3D Printing in Zero-G Experiment, which would put their 3D printer on ISS.

In May 2014, NASA awarded Made In Space a Phase 1 SBIR contract for the development of a recycler unit, to use with the 3D printer on ISS, and for their microwell project. Shortly after, Made In Space was awarded an ISS Space Flight Awareness Award. This award honors "teams that have significantly improved the efficiency, cost or capabilities of space flight."[3]

In June 2014, Made In Space, Inc. showcased their in-space manufacturing capabilities at the White House Maker Faire.[4] The company was nominated as a World Technology Network Award Finalist, in August 2014, in the corporate IT hardware category.[5]

On Sunday, September 21, 2014 at 1:52 a.m. EDT (0552 GMT), Made In Space's Zero-G printer was launched from Cape Canaveral, Florida to ISS, on board Space X CRS-4. On November 17, astronaut Barry "Butch" Wilmore unpacked the 3D printer from its launch packaging. On November 24, at approximately 1:28pm PST, Made In Space successfully printed the first part ever manufactured in space. On December 11, 2014, at Autodesk University 2014, Chef Strategy Officer and co-founder, Mike Chen, revealed Made In Space's first functional application, a buckle developed by NASA astronaut Yvonne Cagle. This buckle is part of exercise equipment to assist with the reduction of muscle loss in zero gravity environments.

On December 12, 2014 the Cooper Hewitt Smithsonian Design Museum re-opened after 3 years of renovation, as the new Cooper Hewitt. Dedicated to contemporary and historic design, Made In Space's 3D printer was one of their grand re-opening exhibits, in their Tools: Extending Our Reach exhibit area.[6] The Zero-G printer was featured along with replicas of 13 of the first 21 objects printed in space and a replica of the plate affixed to the printer, which was created by Jon Lomberg, the artist who designed Voyager's Golden Record.

On December 17, 2014, the first uplinked tool, a ratchet, was manufactured on ISS. Prior to the print of the ratchet, all of the other items manufactured were previously printed on the printer, before it launched, and the files were available via an SD card launched with the printer. The ratchet files were uplinked from the Made In Space office to the ISS space station. The ratchet took four hours to print.[7]

40.2   3D Printing in Zero-G Technology Demonstration

Announced in May 2013, NASA and Made In Space partnered to send the first 3D printer to space, known as the 3D Printing in Zero-G Technology Demonstration (also known as 3D Printing in Zero-G Experiment or 3D Printing in Zero-G). The scientific objective of this experiment is to prove a 3D printer could be developed for use in zero gravity. This experiment "is the first step towards establishing an on-demand machine shop in space, a critical enabling component for deep-space crewed missions and in-space manufacturing."[8]

Integrated into a Microgravity Science Glovebox (MSG), 3D Printing in Zero-G is a proof of concept experiment. It includes printing multiple copies of planned items to test for several variables, including: dimensions, layer adhesion, tensile strength, flexibility and compressional strength. Known as "coupons", these items will be tested by the American Society for Testing and Materials (ASTM) and compared to duplicate items printed on Earth. The comparisons of these space and terrestrial manufactured coupons will be used to further refine future 3D printing in space.[9]

40.2.1   The Jon Lomberg Golden Plate

Inspired by Jon Lomberg's work on the Voyager Golden Record, Lomberg worked with Made In Space to create the Golden Plate, to help commemorate the first manufacturing of something in space. It is attached internally, so it is visible through the front viewing window.

The Golden Plate features imagery that symbolizes both the 3D Printing in Zero-G project, but also the individuals who have been instrumental in bringing the project to fruition. These features include: 27 stars, which represents 16 key contributors from the Marshall Space Flight Center, 10 key Made In Space, Inc. employees and Jon Lomberg, 1 comet to symbolize all of the other people who supported Made In Space and Star Trek character, Jean Luc Picard's catchphrase, "Make it so" in binary, symbolizing the functionality the 3D printer brings to ISS.

40.3   Additive Manufacturing Facility

The Additive Manufacturing Facility (AMF) is scheduled for deployment in 2015. This Nanoracks facility, featuring Made In Space's Zero-G printer, will be used to manufacture parts both for NASA and other space agencies, as needed on ISS. In addition, additive space manufacturing will also be commercially, with the AMF, for anyone around the world.[10]

40.4   Recycling of 3D Printer Plastic

NASA in 2014 selected Made In Space as one of several companies to develop a recycler for 3D printed material on the ISS.[11] In the follow up solicitation Made In Space failed to be selected for the Phase II SBIR.[12][13]

40.5 Dr. Cagle's Buckle

One of the first functional parts to be manufactured in space will be a buckle designed by NASA astronaut Dr. Yvonne Cagle. Used with stretchable material already on ISS, this buckle will be used to gauge lactic acid in muscles, during exercise. By relocating this lactic acid, astronauts will be able to exercise more efficiently and effectively, preventing muscle atrophy and improving cardiovascular health of astronauts. Prior to the ability to manufacture the buckle in space, one of the design concerns was making it strong enough to withstand the forces of launch. Now, the lightweight buckle is designed for functionality, without the launch forces concerns.[14]

40.6 References

[1] "ISS Additive Manufacturing Facility for On-Demand Fabrication in Space". *sbir.gov*. Small Business Innovation Research. Retrieved 27 December 2014.

[2] "ISS Additive Manufacturing Facility for On-Demand Fabrication in Space". *sbir.gov*. Small Business Innovation Research. Retrieved 27 December 2014.

[3] Lowery, Grant. "MADE IN SPACE RECEIVES AMES ISS SPACE FLIGHT AWARENESS AWARD". *MadeInSpace.us*. Made In Space, Inc. Retrieved 27 December 2014.

[4] Pegoraro, Rob. "White House Hosts Its First Maker Faire, with Robotic Giraffe in Attendance". *Yahoo.com*. Yahoo. Retrieved 27 December 2014.

[5] "THE 2014 WORLD TECHNOLOGY AWARD FINALISTS". *wtn.net*. World Technology Network. Retrieved 27 December 2014.

[6] "ABOUT TOOLS: EXTENDING OUR REACH". *cooperhewitt.org*. Cooper Hewitt. Retrieved 27 December 2014.

[7] Afsarifard, Hasti. "THE FIRST UPLINK TOOL MADE IN SPACE IS…". *madeinspace.us*. Made In Space, Inc. Retrieved 27 December 2014.

[8] "3D Printing In Zero-G Technology Demonstration (3D Printing In Zero-G)". *Nasa.gov*. NASA. Retrieved 27 December 2014.

[9] "3D Printing In Zero-G Technology Demonstration (3D Printing In Zero-G)". *nasa.gov*. NASA. Retrieved 27 December 2014.

[10] "PlanetTech News interviews Made In Space". PlanetTech News. Retrieved 8 February 2015.

[11] http://3dprint.com/3559/nasa-made-in-space-3d-print-filament-recycle/

[12] http://sbir.nasa.gov/SBIR/abstracts/14/sbir/phase2/SBIR-14-2-H10.01-9479.html

[13] http://sbir.nasa.gov/prg_selection/node/54501

[14] Thimmesch, Debra. "One of the Next Useful Objects 3D Printed in Space Will Be a Buckle!". *3dprint.com*. 3DPrint.com. Retrieved 27 December 2014.

•

Chapter 41

Magnetically assisted slip casting

Magnetically-assisted slip casting is a manufacturing technique that uses anisotropic stiff nanoparticle platelets in a ceramic, metal or polymer functional matrix to produce[1] layered objects that can mimic natural objects such as nacre. Each layer of platelets is oriented in a different direction, giving the resulting object greater strength. The inventors claimed that the process is 10x faster than commercial 3D printing. The magnetisation and orientation of the ceramic platelets has been patented.[2]

41.1 Technique

The technique involves pouring a suspension of magnetized ceramic micro-plates. Pores in the plaster mold absorb the liquid from the suspension, solidifying the material from the outside in. The particles are subjected to a strong magnetic field as they solidify that causes them to align in one direction. The field's orientation is changed at regular intervals, moving the plates still in suspension, without disturbing already-solidified plates. By varying the composition of the suspension and the direction of the platelets, a continuous process can produce multiple layers with differing material properties in a single object. The resulting objects can closely imitate their natural models.[2]

41.2 Artificial tooth

Researchers produced an artificial tooth whose microstructure mimicked that of a real tooth. The outer layers, corresponding to enamel, were hard and structurally complex. The outer layers contained glass nanoparticles and aluminium oxide plates were aligned perpendicular to the surface. After the outer layers hardened, a second suspension was poured. It contained no glass, and the plates were aligned horizontally to the surface of the tooth. These deeper layers were tougher, resembling dentine. The tooth was then cooked at 1,600 degrees to compact and harden the material — a process known as sintering. The last step involved filling remaining pores with a synthetic monomer used in dentistry, which polymerizes after treatment.[2] Hardness and durability approximated that of both the enamel and dentine of a tooth.[3]

41.3 See also

- 3D printing

- Continuous Liquid Interface Production

41.4 References

[1] Le Ferrand, Hortense; Bouville, Florian; Niebel, Tobias P.; Studart, André R. (2015-09-21). "Magnetically assisted slip casting of bioinspired heterogeneous composites". *Nature Materials*. advance online publication. doi:10.1038/nmat4419. ISSN 1476-4660.

[2] Micu, Alexandru (September 29, 2015). "Artificial tooth is as good as the real deal". ZME. Retrieved 2015-09-29.

[3] Watry,, Greg (September 29, 2015). "Creating Fake Teeth as Strong as the Real Deal". R&D. Retrieved 2015-09-29.

Chapter 42

MatterHackers

MatterHackers is an Orange County-based company founded in 2012 that supplies 3D printing materials and tools.[1] MatterHackers is developing their 3D printer control software, MatterControl.[2]

42.1 History

MatterHackers was founded in 2012, and provides both an online and physical, retail presence for customers.[2] MatterHackers was an exhibitor at the World Maker Faire New York 2013[3] In 2014, MatterHackers will be a sponsor of the 2014 3D Printer World EXPO that's being held in Burbank, California.[4]

42.2 MatterControl

MatterControl is MatterHacker's software for 3D printers. It is in development.[2] "MatterControl is free software for organizing and managing 3D print jobs, with integrated slicing."[5] MatterHackers has stated that while they may provide additional features as paid plug-ins, MatterControl at its core will remain free.[6]

Airwolf 3D offers its customers MatterControl software for their 3D printers.[7]

42.3 Services

MatterHackers also supplies customers with 3D printing goods. In June 2013, MatterHackers opened their own retail location in Lake Forest, California where they sell 3D printing supplies, parts, and accessories. The shop also carries Airwolf and Type A Machines 3D printers.[8]

42.4 See also

- 3D Printing

42.5 References

[1] "Printing in 3D is a snap for this OC business". *Orange County Register*. July 10, 2013. Retrieved November 30, 2013.

[2] Titsch, Mike (July 11, 2013). "MatterHackers Opens 3D Printing Store and Releases MatterControl 0.7.6". *3D Printer World*. Retrieved November 30, 2013.

[3] "MatterHackers". *Maker Faire*. Retrieved November 30, 2013.

[4] "2014 SPONSORS & EXHIBITORS". Retrieved November 30, 2013.

[5] Titsch, Mike (August 30, 2013). "MatterHackers Releases MatterControl 0.8.1". *3D Printer World*. Retrieved November 30, 2013.

[6] "MatterControl FAQ". *MatterControl*. Retrieved November 30, 2013.

[7] "MatterControl Pro Software For Airwolf 3D Printers". *Airwolf3D*. Retrieved November 30, 2013.

[8] Garcia, Eugene (July 11, 2013). "3D printing builds on itself". *Orange County Register*. Retrieved November 30, 2013.

42.6   External links

- Official website
- MatterControl website

Chapter 43

MyMiniFactory

MyMiniFactory is a website dedicated to the free sharing of 3D printable files. The platform is fully curated,[1] which means that every object available on the site has been previously test printed on desktop FDM 3D printers. The object database covers a wide range of different objects sorted in various categories.

43.1 History

MyMiniFactory was launched by iMakr in June 2013[2] and has been growing consistently, crossing the 50,000 mark on Alexa Internet[3] by the end of December 2014.

43.2 Open Source

MyMiniFactory adheres to open source content creation and distribution models. All 3D files are available to freely download, edit and re upload - after going through the validation and test print process. Through this method users are encouraged to create original as well as derivative content, provide feedback and support to other users, and generally stimulate the proliferation of 3D design and 3D printing technology in a free and open manner. In this vein they have partnered with the likes of Oxfam in an attempt to solve humanitarian problems using open source design and manufacture.[4]

43.3 Activity

In addition to sharing files, MyMiniFactory engages with its community of users through various 3D design competitions [5][6] and recently launched MyMiniFactory TV, a streaming platform for 3D designers. MyMiniFactory also innovated through partnerships with various partners such as Royal Mail[7] and Cel Robox, becoming the world's first 3D files library to be integrated to a 3D printer's software.[8]

43.4 References

[1] *Make: 3D Printing: The Essential Guide to 3D Printers By Anna Kaziunas France*, Maker Media, Inc., 19 Nov 2013

[2] *Solidsmack.com*, iMakr Opening My Mini Factory to Provide Free 3D Printable Files (and Sell Yours.)

[3] *Alexa.com*

[4] *policy-practice.oxfam.org.uk*, 3D printing takes emergency response to another level

[5] *3Dprint.com*, MyMiniFactory Sponsoring a 3D Printed GoPro Camera Accessory Design Contest

[6] *http://3dprintingindustry.com/* Win a B9Creator 3D Printer by Winning the Love of Your Life

[7] *ibtimes.co.uk* Royal Mail to trial in-store 3D printing services in London post office

[8] *tctmagazine.com* Robox 3D printer software integrates MyMiniFactorylibrary

43.5 External links

- MyMiniFactory

- MyMiniFactory TV

Chapter 44

Nanophotonic coherent imager

Nanophotonic coherent imagers (NCI) are image sensors that determine both the appearance and distance of an imaged scene at each pixel. It uses an array of LIDARs (scanning laser beams) to gather this information about size and distance, using an optical concept called coherence (wherein waves of the same frequency align perfectly.)[1]

NCIs can capture 3D images of objects with sufficient accuracy to permit the creation of high resolution replicas using 3D printing technology.[1]

The detection of both intensity and relative delay enables applications such as high-resolution 3D reflective and transmissive imaging as well as index contrast imaging.[1]

44.1 Prototype

An NCI using a 4x4 pixel grid of 16 "grating couplers"[2] operates based on a modified time-domain Frequency Modulated Continuous Wave (FMCW) ranging scheme, where concurrent time-domain measurements of both period and the zero-crossing time of each electrical output of the nanophotonic chip allows the NCI to overcome the resolution limits of frequency domain detection.[3] Each pixel on the chip is an independent interferometer that detects the phase and frequency of the signal in addition to the intensity. Each LIDAR pixel spanned only a few hundred microns such that the area fit in area of 300 microns square.[2]

The prototype achieved 15μm depth resolution and 50μm lateral resolution (limited by the pixel spacing) at up to 0.5-meter range. It was capable of detecting a 1% equivalent refractive index contrast at 1mm thickness.[3]

44.2 References

[1] Moss, Richard (April 7, 2015). "New chip could turn phone cameras into high-res 3D scanners". Retrieved April 2015.

[2] "Miniaturized camera chip provides superfine depth resolution for 3D printing". April 6, 2015. Retrieved April 2015.

[3] Aflatouni, Firooz; Abiri, Behrooz; Rekhi, Angad; Hajimiri, Ali (April 2015). "Nanophotonic coherent imager". *Optics Express* **23** (4): 5117–5125. ISSN 1094-4087.

44.3 External links

- New Camera Chip Enables Micrometer-Resolution 3D Images on YouTube

Chapter 45

NovoGen

NovoGen is a proprietary form of 3D printing technology that allows scientists to assemble living tissue cells into a desired pattern. When combined with an extracellular matrix, the cells can be arranged into complex structures, such as organs. Designed by Organovo,[1][2] the NovoGen technology has been successfully integrated by Invetech with a production printer that is intended to help develop processes for tissue repair and organ development.[3][4]

45.1 References

[1] "Structurally and functionally accurate bioprinted human tissue models | Organovo". Organovo. aft. December 9, 2013. Archived from the original on December 31, 2013. Retrieved December 31, 2013. Check date values in: |date= (help)

[2] Forgacs, Andras (September 2013). "Andras Forgacs: Leather and meat without killing animals". *TED: Ideas worth spreading.* Archived from the original on September 22, 2013. Retrieved December 31, 2013.

[3] Wang, Brian (December 18, 2009). "3D Bioprinters". Archived from the original on December 31, 2013. Retrieved December 31, 2013.

[4] "Invetech helps bring bio-printers to life". *Australian Life Scientist.* Westwick-Farrow Media. December 11, 2009. Archived from the original on December 31, 2013. Retrieved December 31, 2013.

Chapter 46

Objet Geometries

Objet is one of the brands of Stratasys, a maker of 3D printers. The brand began with **Objet Geometries Ltd**, a corporation engaged in the design, development, and manufacture of photopolymer 3D printing systems. The company, incorporated in 1998, was based in Rehovot, Israel. In 2011 it merged with Stratasys. It held patents on a number of associated printing materials that are used in *PolyJet* and *PolyJet Matrix* polymer jetting technologies. It distributed 3D printers worldwide through wholly owned subsidiaries in the United States (**Objet Geometries Inc**), Europe (**Objet Geometries GmbH**), and Hong Kong. Objet Geometries owned more than 50 patents and patent-pending inventions.

46.1 History

Objet was founded in 1998 by Rami Bonen, Gershon Miller and Hanan Gotaiit. In September 2000 it announced the completion of a second private placement, securing it $15 million at a post-money company value of $36 million. Participants in this round were the Templeton Foreign Fund, private investors from Europe and the United States and Scitex Corporation, which acquired an initial 18.7% stake in the company, which was subsequently increased.[1] In June 2005, Scitex sold all its interest, then standing at 22.9%, to the other shareholders of Objet for $3.0 million in cash.[2]

46.1.1 Merger with Stratasys

On April 16, 2012 Objet announced that it agreed to merge with Stratasys, a leading manufacturer of 3D printers; in an all-stock transaction. Stratasys shareholders were expected to own 55 percent and Objet shareholders were expected to own 45 percent of the combined company. The merger was completed on December 3, 2012 the market capitalization of the new company was approximately $3.0 billion.[3]

46.2 Technology

The Polyjet matrix 3D printing technology uses simultaneous jetting of multiple types of modeling materials to create a single piece 3D model. PolyJet is used by automotive, electronics, consumer goods, medical development, and clothing manufacturers, as well as for creating 3D models for use in movies such as Coraline.[4]

The Eden line of 3D Printing Systems and the Alaris30 3D desktop printer are based on the PolyJet technology. The Connex family of 3D printers is based on PolyJet Matrix technology, which jets multiple model materials simultaneously and creates composite *Digital Materials* on the fly. All Objet systems use FullCure polymers.

46.3 Patents

Objet owns more than 50 patents and patent pending inventions:

- 3D Printing Patent - Example 1
- 3D Printing Patent - Example 2
- 3D Printing Patent - Example 3

46.4 Partnerships

Objet Geometries has partnered with SolidWorks to interface their computer-aided design software with Objet's Connex500 system. The co-developed software add-in allows significantly more control over end to end modelling preferences.

46.5 Awards

Frost & Sullivan recognized Object Geometries in 2008 with their Annual Excellence Award [5]

In 2009, Objet's Alaris desktop 3D printer was recognized with the Plastpol Award at the annual plastics show in Poland.[6]

46.6 See also

- 3D printing
- Rapid prototyping
- Stereolithography

46.7 Notes

[1] Scitex Announces Investments in Objet Geometries, BUSINESS WIRE -June 28, 2000

[2] Scitex to Sell its Holdings in Objet Geometrie, PRNewswire, 24 May 2005

[3] Stratasys and Objet Complete Merger; BusinessWire, December 3, 2012

[4] http://www.hk3dprinting.co.uk/stereolithography-sla.html

[5] Frost & Sullivan Honors Object Geometries with Annual Excellence Award

[6] for Alaris30 Plastpol 2009.pdf Plastpol Award

46.8 References

- Frank W. Liou (2008). *Rapid prototyping and engineering applications: a toolbox for prototype development*. Boca Raton: CRC Press. ISBN 0-8493-3409-8.

- Andreas Gebhardt (2003). *Rapid Prototyping*. Hanser Verlag. ISBN 978-1-56990-281-3.

- Paulo Jorge Bártolo (2007). *Virtual and Rapid Manufacturing*. Routledge. ISBN 978-0-203-93187-5.

- Weiyin Ma; Patri K. Venuvinod (2003). *Rapid Prototyping: Laser-Based and Other Technologies*. Berlin: Springer. ISBN 1-4020-7577-4.

46.9 External links

- Official website

- Independent review of the Polyjet Matrix Technology

- List of Patent by Objet at Google Patents

Chapter 47

Pinshape

Pinshape Inc. is an online 3D printing community and marketplace with headquarters in Vancouver, Canada.[1][2] It allows designers to share and sell their 3D printable designs, and people with 3D printers to print those designs on their own printers.[3][4][5]

47.1 About

Pinshape was founded in 2013 by Lucas Matheson (CEO), Nick Schwinghamer (COO), and Andre Yanes (CTO).[6] The site is a marketplace that showcases the digital work of 3D designers from all over the world. 3D print designers set their own prices for their design files, and also choose which license to offer their work under (Creative Commons or other). People with 3D printers can browse the selection of designs and then either get the file for free to print themselves, or pay the designer for access to the file before printing.

Designs found on Pinshape can be directly downloaded if the designer allows, or they can be sent directly to a user's 3D printer using a direct browser-to-printer experience that removes the need to access the design source file and thus, increases intellectual property (IP) security.[1] Utilizing a cloud slicing and file streaming technology, designers have the option of charging per print, so that 3D files aren't stored on a customer's computer. Pinshape also allows it's users to review designs and share the settings they used to print off the files.[1]

47.2 500 Startups

Pinshape was selected for and attended the 500 Startups accelerator program in Mountain View, California as part of Batch 9, from April to July 2014. They were one of 30 companies selected from over 1400 applicants to participate in the 4-month program.[7][8]

47.3 References

[1] "Secure Streaming Tech Reduces IP Risks for Designers". Retrieved 2015-05-07.

[2] "Building a Simple 3D Print Experience via Partnerships". *tech.co*. Retrieved 2015-05-07.

[3] "CES Comes to the Capitol". *tech.co*. Retrieved 2015-05-09.

[4] "How 3D Printing is Fueling the Explosion of Open Maker Marketplaces". 2014-10-14. Retrieved 2015-05-09.

[5] "Meet the 'iTunes of 3D Printing' That Helps Designers Get Paid". Retrieved 2015-05-09.

[6] "Pinshape Infograph & Survey: Who's 3D Designing and Printing?". Retrieved 2015-05-11.

[7] Lawler, Ryan. "500 Startups Accelerator Announces Its Ninth Batch Of Companies And Two New EIRs". *TechCrunch*. Retrieved 2015-05-21.

[8] "Two More Canadian Startups in Store for 500 Startups Awesomeness - Techvibes.com". *www.techvibes.com*. Retrieved 2015-05-21.

Chapter 48

PLY (file format)

PLY is a computer file format known as the **Polygon File Format** or the **Stanford Triangle Format**.

The format was principally designed to store three-dimensional data from 3D scanners. It supports a relatively simple description of a single object as a list of nominally flat polygons. A variety of properties can be stored including: color and transparency, surface normals, texture coordinates and data confidence values. The format permits one to have different properties for the front and back of a polygon.

There are two versions of the file format, one in ASCII, the other in binary.

48.1   The File Format

A complete description of the PLY format is beyond the scope of this article - but one may obtain a good understanding of the basic concepts from the following description:

Files are organised as a header, that specifies the elements of a mesh and their types, followed by the list of elements itself, usually vertices and faces - potentially other entities such as edges, samples of range maps, and triangle strips can be encountered.

The header of both ASCII and binary files is ASCII text. Only the numerical data that follows the header is different between the two versions.

The header always starts with a "magic number", a line containing

ply

which identifies the file as a PLY file. The second line indicates which variation of the PLY format this is. It should be one of:

format ascii 1.0 format binary_little_endian 1.0 format binary_big_endian 1.0

Future versions of the standard will change the revision number at the end - but 1.0 is the only version currently in use.

Comments may be placed in the header by using the word comment at the start of the line. Everything from there until the end of the line should then be ignored. e.g.:

comment This is a comment!

The 'element' keyword introduces a description of how some particular data element is stored and how many of them there are. Hence, in a file where there are 12 vertices, each represented as a floating point (X,Y,Z) triple, one would expect to see:

element vertex 12 property float x property float y property float z

Other 'property' lines might indicate that colours or other data items are stored at each vertex and indicate the data type of that information. Regarding the data type there are two variants, depending on the source of the ply file, the type can

177

be specified with one of *char uchar short ushort int uint float double*, or one of *int8 uint8 int16 uint16 int32 uint32 float32 float64*. For an object with ten polygonal faces, one might see:

element face 10 property list uchar int vertex_indices

The word 'list' indicates that the data is a list of values–the first of which is the number of entries in the list (represented as a 'uchar' in this case) and each list entry is (in this case) represented as an 'int'.

At the end of the header, there must always be the line:

end_header

48.2   ASCII or Binary Format

In the ASCII version of the format, the vertices and faces are each described one to a line with the numbers separated by white space. In the binary version, the data is simply packed closely together at the 'endianness' specified in the header and with the data types given in the 'property' records. For the common "property list..." representation for polygons, the first number for that element is the number of vertices that the polygon has and the remaining numbers are the indices of those vertices in the preceding vertex list.

48.3   History

The PLY format was developed in the mid-90s by Greg Turk and others in the Stanford graphics lab under the direction of Marc Levoy. Its design was inspired by the Wavefront .obj format, but the Obj format lacked extensibility for arbitrary properties and groupings, so the "property" and "element" keywords were devised to generalize the notions of vertices, faces, associated data, and other groupings.

48.4   See also

- STL (file format)

- Additive Manufacturing File Format

- Wavefront .obj file, a 3D geometry definition file format with *.obj* file extension

- MeshLab: an open source Windows, Mac OS X and Linux application for visualizing, processing and converting three-dimensional meshes to or from the PLY file format.

- CloudCompare, another open source application for handling PLY files.

- Mathematica A technical computing system that can work with PLY files.

48.5   External links

- PLY - Polygon File Format

- Some tools for working with PLY files (C source code)

- rply - An Ansi C software library for reading and writing PLY files (MIT license)

- libply - A C++ software library for reading and writing PLY files (GNU license)

- Another C++ software library for reading and writing PLY files (GPL 3.0 license)

- A repository of 3D models stored in the PLY format

Chapter 49

Polyphenylsulfone

Polyphenylsulfone (PPSF or PPSU) is a type of moldable plastic often used in rapid prototyping and rapid manufacturing (direct digital manufacturing) applications. Polyphenylsulfone is a heat and chemical-resistant material. It is typically used in automotive, aerospace and plumbing applications. Polyphenylsulfone has virtually no melting point, due to its amorphous nature,[1] and offers tensile strength up to 55 MPa (8000 psi). Its commercial name is **RadelR**. In plumbing applications, ployphenylsulfone fittings have been found to sometimes form cracks prematurely or to experience failure when improperly installed using non-manufacturer approved installation methods or systems.[2]

49.1 References

[1] PPSF for FORTUS 3D Production Systems

[2] Failure Analysis of Plastic Crimp Fitting Assemblies

Chapter 50

Powder bed and inkjet head 3D printing

This article is about powder bed and inkjet-based 3d printing. For the popular term for all additive manufacturing processes, see additive manufacturing.

Powder bed and inkjet 3d printing, known variously as "binder jetting" and "drop-on-powder" – or simply "3d printing" (3DP) – is a rapid prototyping and additive manufacturing (or "layered manufacturing") technology for making objects described by digital data. (Other "powder-bed" manufacturing technologies include Selective Laser Sintering and Selective Laser Melting.)

50.1 History

This technology was first developed at the Massachusetts Institute of Technology in 1993 and in 1995 Z Corporation obtained an exclusive license. The term "Three-Dimensional Printing" was trademarked by the same.[1][2]

50.2 Description

As in many other additive manufacturing processes, (and as a layered manufacturing technology), the part to be printed is built up from many thin cross sections of the 3D model. An inkjet print head moves across a bed of powder, selectively depositing a liquid binding material. A thin layer of powder is spread across the completed section and the process is repeated with each layer adhering to the last.

When the model is complete, unbound powder is automatically and/or manually removed in a process called "de-powdering" and may be reused to some extent.[3]

The de-powdered part could optionally be subjected to various infiltrants or other treatments to produce properties desired in the final part.

50.3 Materials

In the original implementations, starch and gypsum plaster fill the powder bed, the liquid "binder" being mostly water to activate the plaster. The binder also includes dyes (for color printing), and additives to adjust viscosity, surface tension, and boiling point to match print head specifications. The resulting plaster parts typically lack "green strength" and require infiltration by melted wax, cyanoacrylate glue, epoxy, etc. before regular handling.

While not necessarily employing conventional inkjet technology, various other powder-binder combinations may be deployed to form objects by chemical or mechanical means. The resulting parts may then be subjected to different post-processing regimes, such as infiltration or bakeout. This may be done, for example, to eliminate the mechanical binder (e.g., by burning) and consolidate the core material (e.g., by melting), or to form a composite material blending the properties of powder and binder. Depending on the material, full color printing may or may not be an option. As of 2014, inventors and manufacturers have developed systems for forming objects from sand and calcium carbonate (forming a synthetic marble), acrylic powder and cyanoacrylate, ceramic powder and a liquid binder, sugar and water (for making candies), etc.

3D printing technology has a limited potential to vary material properties in a single build, but is generally limited by the use of a common core material. In the original Z Corporation systems, cross-sections are typically printed with solid outlines (forming a solid shell) and a lower-density interior pattern to speed printing and ensure dimensional stability as the part cures.

50.4   Characteristics

In addition to volumetric color by use of multiple print heads and colored binder, the 3D printing process is generally faster than other additive manufacturing technologies such as Fused Deposition Manufacturing or drop-on-drop material jetting which require 100% of build and support material to be deposited at the desired resolution. In 3D printing, the bulk of each printed layer, regardless of complexity, is deposited by the same, rapid spreading process.[4]

As with other powder-bed technologies, support structures are generally not required because loose powder supports overhanging features and stacked or suspended objects. The elimination of printed support structures can reduce build time and material use and simplify both equipment and post-processing. However, de-powdering itself can be a delicate, messy, and time-consuming task. Some machines therefore automate de-powdering and powder recycling to what extent feasible. Since the entire build volume is filled with powder, as with stereolithography, means to evacuate a hollow part must be accommodated in the design.

Like other powder-bed processes, surface finish and accuracy, object density, and—depending on the material and process—part strength may be inferior to technologies such as stereolithography (SLA) or Selective laser sintering (SLS). Although "stair-stepping" and asymmetrical dimensional properties are features of 3D printing as most other layered manufacturing processes, 3D printing materials are generally consolidated in such a way that minimizes the difference between vertical and in-plane resolution. The process also lends itself to rasterization of layers at target resolutions, a fast process that can accommodate intersecting solids and other data artifacts.

Powder bed and inkjet 3D printers are expensive compared to regular 3D printers with prices ranging from $50.000 to $2 Million for enterprise grade, but recent $1.300 model introduced by Yvo de Haas and released as open source project make it more affordable.[5]

50.5   See also

- 3D printing

- List of common 3D test models

- 3D printing marketplace

- Volumetric printing

50.6   References

[1] "Printers produce copies in 3D". *BBC News*. August 6, 2003. Retrieved October 31, 2008.

[2] Grimm, Todd (2004). *User's Guide to Rapid Prototyping*. SME. p. 163. ISBN 978-0-87263-697-2. Retrieved October 31, 2008.

[3] Sclater, Neil; Nicholas P. Chironis (2001). *Mechanisms and Mechanical Devices Sourcebook.* McGraw-Hill Professional. p. 472. ISBN 978-0-07-136169-9. Retrieved October 31, 2008.

[4] http://www.fusion3design.com/

[5] http://3dprint.com/12560/plan-b-3dp-3d-printer/

Chapter 51

Print the Legend

Print the Legend is a 2014 documentary film and Netflix Original focused on the 3D printing revolution.[1] It delves into the growth of the 3D printing industry, with focus on companies MakerBot, Formlabs, Stratasys, and 3D Systems, as well as figures of controversy in the industry such as Cody Wilson.

The title of the film comes from the denouement of the film *The Man Who Shot Liberty Valance*.[2]

A trailer for the film is available at the documentary's official website, http://printthefilm.com.

51.1 Plot Summary

Print the Legend portrays some of the 3D printing histories and achievements of several 3D printing companies, including MakerBot, Formlabs, Stratasys, and 3D Systems.

The documentary also explores the relationship between the 3D printing industry and the gun rights advocacy movement. In particular, Cody Wilson, who is known for gun rights advocacy and specifically for promoting the 3D printing of guns, is interviewed extensively in the documentary.[3]

51.2 Cast

- Chris Anderson
- Bruce Bradshaw
- Craig Broady
- Bill Buel
- Michael Calore
- Nadia Cheng
- Alan Cramer
- David Cranor
- Michael Curry
- Malo Delarue
- Brad Feld

- Ian Ferguson
- Lorenzo Franceschi-Bicchierai
- Martin Galese
- Matt Griffin
- James Gunipero
- Zach Hoeken
- Luke Iseman
- Annelise Jeske
- Brad Kenney
- Eric Klein
- Cliff Kuang
- Jenny Lawton
- Natan Linder
- Ira Liss
- Alex Lobovsky
- Larisa Lobovsky
- Maxim Lobovsky
- Marty Markowitz
- Adam Mayer
- Nathan Meyers
- Jennifer Milne
- Anthony Moschella
- Will O'Brien
- Jeff Osborn
- Andrew Pelkey
- Bre Pettis
- Chuck Pettis
- Yoav Reches
- Avi Reichental
- David Reis
- Barry Schuler
- Virginia White
- Cody Wilson
- Luke Winston

51.3   Festivals

Awards

- Special Jury Recognition Award - SXSW Film Festival (2014)[1]

- Special Jury Award - IFF Boston (2014)[1]

51.4   References

[1] "Print the Legend". *Print the Legend*. Retrieved 9 September 2014.

[2] Luis Lopez and Steven Klein interviewed on the TV show Triangulation on the TWiT.tv network

[3] http://www.nytimes.com/2014/09/26/movies/print-the-legend-looks-at-3-d-printers.html?_r=0

51.5   External links

- *Print the Legend* at the Internet Movie Database

Chapter 52

Projection micro-stereolithography

Projection micro-stereolithography (PμSL) adapts 3D printing technology for micro-fabrication. Digital micro display technology provides dynamic stereolithography masks that work as a virtual photomask. This technique allows for rapid photopolymerization of an entire layer with a flash of UV illumination at micro-scale resolution. The mask can control individual pixel light intensity, allowing control of material properties of the fabricated structure with desired spatial distribution.

Materials include polymers, responsive hydrogels, shape memory polymers and bio-materials.[1]

52.1 Process

The dynamic mask defines the beam. The beam is focused on the surface of a UV-curable polymer resin through a projection lens that reduces the image to the desired size. Once a layer is polymerized, the stage drops the substrate by a predefined layer thickness, and the dynamic mask displays the image for the next layer on top of the preceding one. This proceeds iteratively until complete. The process can create layer thickness on the order of 400 nm.[2]

Sub 2 μm horizontal and sub-1 μm vertical resolutions have been achieved, with sub-1 μm feature sizes. Process can work at ambient temperature and atmosphere, although increased nitrogen improves polymerization Production rates of 4 cu mm/hr have been achieved, depending on resin viscosity.[2]

Materials can be easily switched during fabrication, enabling integration of multiple material elements in a single process.[2]

52.2 Applications

Applications include creating molds, electroplating or (with resin additives) ceramic items, including micro-bio reactors to support tissue growth, micromatrices for drug delivery and detection and biochemical integrated circuits to simulate biological systems.[2]

52.3 See also

- Continuous Liquid Interface Production

52.4 References

[1] "Projection Micro-Stereolithography". MIT Department of Mechanical Engineering. Retrieved April 2015.

[2] Fang, Nicholas. "Projection Microstereolithography" (PDF). Department of Mechanical Science & Engineering, University of Illinois. Retrieved April 2015.

Chapter 53

Proto BuildBar

Proto BuildBar is a privately-owned commercial makerspace and café located in Dayton, Ohio. It is possibly the first public hakerspace/restaurant to open in the United States.[1]<ref name="2News article">Edmé, Beairshelle (1 November 2014). "New business merges tech and drinks at cafe bar". *WDTN*. Retrieved 2 March 2015.</ref>[2]

53.1 References

[1] Cogliano, Joe (10 October 2014). "New downtown Dayton spot combines technology, trendy cafe". *Dayton Business Journal* (Dayton, Ohio). Retrieved 17 October 2014.

[2] Barrow, Olivia (24 October 2014). "Downtown Dayton turns a corner". *Dayton Business Journal* (Dayton, Ohio). Retrieved 2 March 2015.

53.2 External links

- Proto BuildBar website

- Proto BuildBar covered by Laughing Squid

- Proto BuildBar covered by Hackaday

- Proto BuildBar covered by 3DPrint.com

Chapter 54

Rapid prototyping

This article is about rapid prototyping of physical objects. For rapid software prototyping, see rapid application development.

Rapid prototyping is a group of techniques used to quickly fabricate a scale model of a physical part or assembly using

A rapid prototyping machine using selective laser sintering

three-dimensional computer aided design (CAD) data.[1][2] Construction of the part or assembly is usually done using 3D printing or "additive layer manufacturing" technology.[3]

The first methods for rapid prototyping became available in the late 1980s and were used to produce models and prototype

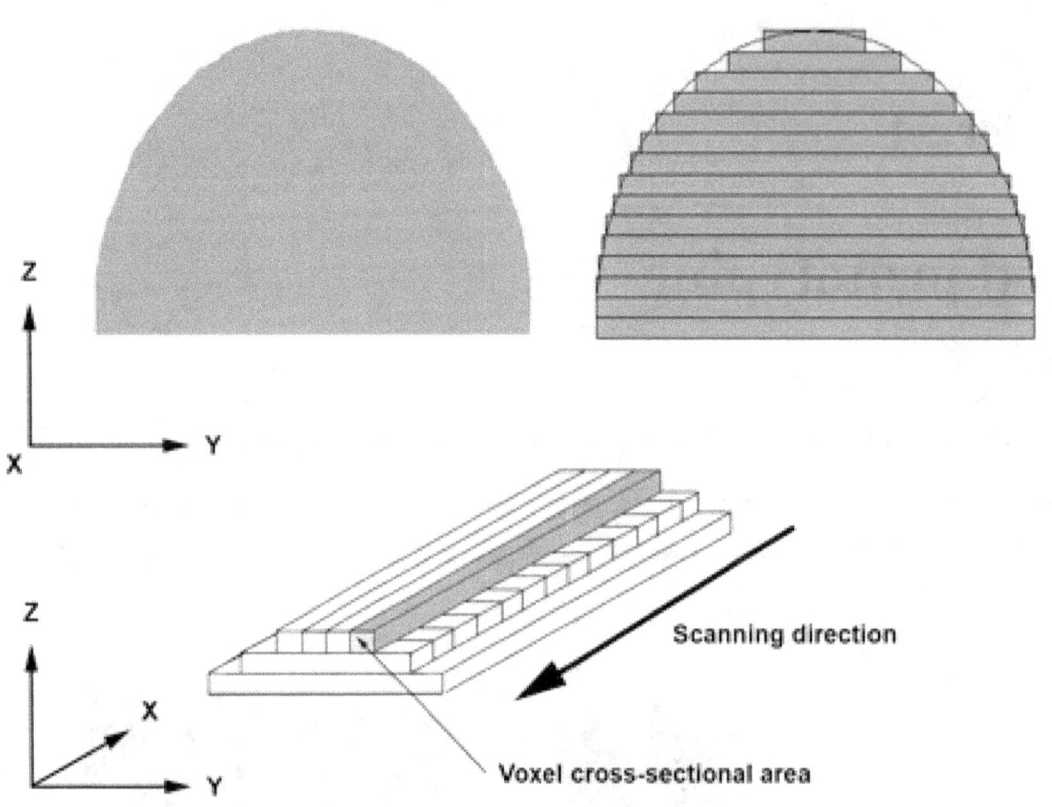

3D model slicing

parts. Today, they are used for a wide range of applications[4] and are used to manufacture production-quality parts in relatively small numbers if desired without the typical unfavorable short-run economics. This economy has encouraged online service bureaus. Historical surveys of RP technology[2] start with discussions of simulacra production techniques used by 19th-century sculptors. Some modern sculptors use the progeny technology to produce exhibitions.[5] The ability to reproduce designs from a dataset has given rise to issues of rights, as it is now possible to interpolate volumetric data from one-dimensional images.

As with CNC subtractive methods, the computer-aided-design - computer-aided manufacturing CAD-CAM workflow in the traditional Rapid Prototyping process starts with the creation of geometric data, either as a 3D solid using a CAD workstation, or 2D slices using a scanning device. For RP this data must represent a valid geometric model; namely, one whose boundary surfaces enclose a finite volume, contain no holes exposing the interior,and do not fold back on themselves. In other words, the object must have an "inside." The model is valid if for each point in 3D space the computer can determine uniquely whether that point lies inside, on, or outside the boundary surface of the model. CAD post-processors will approximate the application vendors' internal CAD geometric forms (e.g., B-splines) with a simplified mathematical form, which in turn is expressed in a specified data format which is a common feature in Additive Manufacturing: STL (stereolithography) a de facto standard for transferring solid geometric models to SFF machines. To obtain the necessary motion control trajectories to drive the actual SFF, Rapid Prototyping, 3D Printing or Additive Manufacturing mechanism, the prepared geometric model is typically sliced into layers, and the slices are scanned into lines [producing a "2D drawing" used to generate trajectory as in CNC`s toolpath], mimicking in reverse the layer-to-layer physical building process.[2]

54.1 Rapid prototyping and production automotive spareparts

Electric cars can be built and tested in one year with 3D production systems. [6]

54.2 History

In the 1970s, Joseph Henry Condon and others at Bell Labs developed the Unix Circuit Design System (UCDS), automating the laborious and error-prone task of manually converting drawings to fabricate circuit boards for the purposes of research and development.

In the 1980s U.S. policy makers and industrial managers were forced to take note that America's dominance in the field of machine tool manufacturing evaporated, in what was named the machine tool crisis. Numerous projects sought to counter these trends in the traditional CNC CAM area, which had begun in the US. Later when Rapid Prototyping Systems moved out of labs to be commercialized it was recognized that developments were already international and U.S. rapid prototyping companies would not have the luxury of letting a lead slip away. The National Science Foundation was an umbrella for the National Aeronautics and Space Administration (NASA), the US Department of Energy, the US Department of Commerce NIST, the US Department of Defense, Defense Advanced Research Projects Agency (DARPA), and the Office of Naval Research coordinated studies to inform strategic planners in their deliberations. One such report was the 1997 Rapid Prototyping in Europe and Japan Panel Report[2] in which Joseph J. Beaman[7] founder of DTM Corporation [DTM RapidTool pictured] provides a historical perspective : The roots of rapid prototyping technology can be traced to practices in topography and photosculpture. Within TOPOGRAPHY Blanther (1892) suggested a layered method for making a mold for raised relief paper topographical maps .The process involved cutting the contour lines on a series of plates which were then stacked. Matsubara (1974) of Mitsubishi proposed a topographical process with a photo-hardening photopolymer resin to form thin layers stacked to make a casting mold. PHOTOSCULPTURE was a 19th-century technique to create exact three-dimensional replicas of objects. Most famously Francois Willeme (1860) placed 24 cameras in a circular array and simultaneously photographed an object.The silhouette of each photograph was then used to carve a replica. Morioka (1935, 1944) developed a hybrid photo sculpture and topographic process using structured light to photographically create contour lines of an object.The lines could then be developed into sheets and cut and stacked, or projected onto stock material for carving. The Munz(1956) Process reproduced a three-dimensional image of an object by selectively exposing, layer by layer, a photo emulsion on a lowering piston. After fixing, a solid transparent cylinder contains an image of the object. [8]

The technologies referred to as Solid Freeform Fabrication are what we recognize today as Rapid Prototyping, 3D Printing or Additive Manufacturing: Swainson (1977), Schwerzel (1984) worked on polymerization of a photosensitive polymer at the intersection of two computer controlled laser beams. Ciraud (1972) considered magnetostatic or electrostatic deposition with electron beam, laser or plasma for sintered surface cladding. These were all proposed but it is unknown if working machines were built. Hideo Kodama of Nagoya Municipal Industrial Research Institute was the first to publish an account of a solid model fabricated using a photopolymer rapid prototyping system (1981).[2] Even at that early date the technology was seen as having a place in manufacturing practice. A low resolution, low strength output had value in design verification, mould making, production jigs and other areas. Outputs have steadily advanced toward higher specification uses.[9]

Innovations are constantly being sought,to improve speed and the ability to cope with mass production applications.[10] A dramatic development which RP shares with related CNC areas is the freeware open-sourcing of high level applications which constitute an entire CAD-CAM toolchain. This has created a community of low res device manufacturers. Hobbyists have even made forays into more demanding laser-effected device designs.[11]

54.3 See also

- Digital modeling and fabrication

- Fab lab

- Laser engineered net shaping

- Open hardware

- Von Neumann universal constructor

54.4 References

[1] eFunda, Inc. "Rapid Prototyping: An Overview". Efunda.com. Retrieved 2013-06-14.

[2] NSF JTEC/WTEC Panel Report-RPA http://www.wtec.org/pdf/rp_vi.pdf

[3] "Interview with Dr Greg Gibbons, Additive Manufacturing, WMG, University of Warwick", Warwick University, Knowledge-Centre. Accessed 18 October 2013

[4] medical applications of rapid prototyping intech open books http://cdn.intechopen.com/pdfs/20116/InTech-medical_applications_of_rapid_prototyping_a_new_horizon.pdf

[5] sculpture exhibition School of the Art Institute of Chicago http://blogs.saic.edu/sugs/exhibitions/artifact/

[6] Revolutionary New Electric Car Built and Tested in One Year with Objet1000 Multi-material 3D Production System

[7] history of laser Additive Manufacturing http://www.lia.org/blog/2012/04/the-history-of-laser-additive-manufacturing/

[8] JTEC/WTEC Panel Report on Rapid Prototyping in Europe and Japan pg.24

[9] SME Wolhers/

[10] Hayes, Jonathan (2002) Concurrent printing and thermographing for rapid manufacturing: executive summary. EngD thesis, University of Warwick.. Accessed 18 October 2013

[11] "Will 3D Printing Push Past the Hobbyist Market?", Fiscal Times, 2 September 2013. Accessed 18 October 2013

54.5 Bibliography

- Wright, Paul K. (2001). *21st Century Manufacturing*. New Jersey: Prentice-Hall Inc.

54.6 External links

- Rapid prototyping websites at DMOZ

Chapter 55

Recyclebot

A **recyclebot** (or RecycleBot) is an open-source hardware device for converting waste plastic into filament for open-source 3D printers like the RepRap.[1] Making DIY 3D printer filament at home is both less costly and better for the environment than purchasing conventional 3D printer filament.[2][3][4]

55.1 Recyclebot technology

RepRap 3D printers have been shown to reduce costs for consumers by offsetting purchases that can be printed.[5][6][7][8] The RepRap's plastic feedstock is one area where cost can still be reduced. In 2014 professor Joshua Pearce pointed out that "Filament is retailing for between \$36 and \$50 a kilogram and you can produce your own filament for 10 cents a kilogram if you use recycled plastic"[9] The device can thus further enhance RepRap affordability by reducing operating costs.[10] In addition, to assisting prosumers to reduce their reliance on purchased products, following an open source model, the RepRap and the recyclebot, have made it feasible for 3D printing to be used for small-scale manufacturing to aid sustainable development.[11][12]

The RecycleBot is an open-source hardware project – thus its plans are freely available on the Internet.

- RecycleBot curated by academics in Canada and the U.S. on Appropedia (here) and at the RepRap Wiki (here). For example, the full parts list (or bill of materials for the metal and electronic components and the controls are available on Thingiverse.[13][14]

- Lyman Filament Extruder – a DIY recyclebot

It has been postulated that recycled filament production could also offer an alternative income source by the Ethical Filament Foundation[15][16] or as a form of "fair trade filament".[17]

55.2 History

The history of the RecycleBot was largely derived from the work on the RepRap Wiki under GNU Free Documentation License1.2.[18]

The first recyclebot was developed by university students in Australia.[19] This design was a proof of concept and was a hand-powered design, which thus had a good environmental or ecological footprint did not create filament of high enough quality to be useful for 3D printers. The design for the waste plastic extruder (Recyclebot v2.0 and v2.1) developed at Queen's University Canada and Michigan Tech was heavily influenced by the Web4Deb extruder, which extrudes HDPE for use as a growth medium in aquaponics.[20] This design for the recyclebot was developed, tested and published in the peer-reviewed rapid prototyping literature.[21] This device proved viable for producing 3D printing filament. The

Recyclebot v2.2 is now being carried out by the Michigan Tech in Open Sustainability Technology Research Group.[22] Many makers or DIY enthusiasts have made various versions of RecycleBots, with the most notable being the Lyman Filament Extruder as Lyman, a retired engineer won a design contest to make a low-cost 3D filament fabrication system.[23] There are now many types of recyclebots, many of which are at the early stages of commercialization (in 2014).

55.3 Commercialization

Several versions of open-source RecycleBots have been commercialized via crowd funding such as a with Kickstarter that include:

- Filastruder [24]
- Filafab[25]
- Filabot[26]

55.4 Futurist speculation

It has been hypothesized that such recycling with recyclebots and distributed production with 3D printing will lead to a zero marginal cost society by Jeremy Rifkin.[27] The science-fiction author, Bruce Sterling also wondered in Wired if recyclebots and 3D printers might be used to turn waste into guns.[28] Recyclebots can provide a new method of recycling.[29]

55.5 References

[1] Christian Baechler, Matthew DeVuono, and Joshua M. Pearce, "Distributed Recycling of Waste Polymer into RepRap Feedstock" *Rapid Prototyping Journal,* **19**(2), pp. 118–125 (2013).free open access copy

[2] M.A. Kreiger, M.L. Mulder, A.G. Glover, J. M. Pearce, Life Cycle Analysis of Distributed Recycling of Post-consumer High Density Polyethylene for 3-D Printing Filament, *Journal of Cleaner Production*, DOI:http://dx.doi.org/10.1016/j.jclepro.2014.02.009. open access

[3] The importance of the Lyman Extruder, Filamaker, Recyclebot and Filabot to 3D printing – VoxelFab, 2013.

[4] M. Kreiger, G. C. Anzalone, M. L. Mulder, A. Glover and J. M Pearce (2013). Distributed Recycling of Post-Consumer Plastic Waste in Rural Areas. MRS Online Proceedings Library, 1492, mrsf12-1492-g04-06 doi:10.1557/opl.2013.258. open access

[5] B.T. Wittbrodt, A.G. Glover, J. Laureto, G.C. Anzalone, D. Oppliger, J.L. Irwin, J.M. Pearce (2013), Life-cycle economic analysis of distributed manufacturing with open-source 3-D printers, *Mechatronics*, 23 (2013), pp. 713–726. open access

[6] Study: At-home 3-D printing could save consumers 'thousands' – CNN, 2013

[7] Printing Keychains and Shower Heads: 3-D Printing Goes Beyond the Lab – ABC News

[8] A 3-D Printer Can Pay For Itself In Less Than A Year – Popular Science, 2013

[9] Turning old plastic into 3D printer filament is greener than conventional recycling – 3Ders, 2014

[10] Study: At-home 3-D printing could save consumers 'thousands' – CNN, 2013

[11] 3-D Printing of Open Source Appropriate Technologies for Self-Directed Sustainable Development

[12] DJ Pangburn. 2014.How 3D Printers Are Boosting Off-The-Grid, Underdeveloped Communities - MotherBoard

[13] http://www.thingiverse.com/thing:12948

[14] http://www.thingiverse.com/thing:54180

[15] http://techfortrade.org/our-initiatives/3d4d-challenge/the-ethical-filament-foundation/ Tech for Trade – Ethical Filament Foundation

[16] Charity Targets 3D Printing's Plastic Waste Problem With Standards For An Ethical Alternative 7 November 2013 by Natasha Lomas, Tech Crunch, http://techcrunch.com/2013/11/07/ethical-additive-manufacturing/

[17] Feeley, S. R., Wijnen, B., & Pearce, J. M. (2014). Evaluation of Potential Fair Trade Standards for an Ethical 3-D Printing Filament. *Journal of Sustainable Development*, **7**(5), 1-12. DOI: 10.5539/jsd.v7n5p1

[18] http://reprap.org/wiki/Recyclebot

[19] Recyclebot v1.0

[20] Web4Deb's blog.

[21] Christian Baechler, Matthew DeVuono, and Joshua M. Pearce, "Distributed Recycling of Waste Polymer into RepRap Feedstock" *Rapid Prototyping Journal*, **19**(2), pp. 118–125 (2013).free open access copy

[22]

[23] Harry McCracken (4 March 2013). "How an 83-Year-Old Inventor Beat the High Cost of 3D Printing". Time.

[24] Filastruder

[25] Filafab

[26] Filabot

[27] Jeremy Rifkin, Zero Marginal Cost Society, Palgrave Macmillan, 2014.

[28] 3D Printed gun moving from sinister joke to sinister business model By Bruce Sterling – Wired – Beyond the Beyond

[29] Baltodano, S. (2013). RISE. http://www.mme.fiu.edu/wp-content/uploads/2013/12/F13-OR-T-4.pdf

Chapter 56

RepRap Project

The **RepRap project** started as a British initiative to develop a 3D printer that can print most of its own components and be a low-cost 3D printer, but it is now made up of hundreds of collaborators world wide.[1] RepRap (short for *replicating rapid prototyper*) uses an additive manufacturing technique called *fused filament fabrication* (FFF) to lay down material in layers; a plastic filament or metal wire is unwound from a coil and supplies material to produce a part. The project calls it *Fused Filament Fabrication* (FFF) to avoid trademark issues with the "fused deposition modeling" term.

As an open design, all of the designs produced by the project are released under a free software license, the GNU General Public License.[2]

Due to the self-replicating ability of the machine, authors envision the possibility to cheaply distribute RepRap units to people and communities, enabling them to create (or download from the Internet) complex products without the need for expensive industrial infrastructure (distributed manufacturing)[3] including scientific equipment.[4][5] They intend for the RepRap to demonstrate evolution in this process as well as for it to increase in number exponentially.[1][6] A preliminary study has already shown that using RepRaps to print common products results in economic savings, which justifies the investment in a RepRap.[7]

56.1   History

RepRap was founded in 2005 by Dr Adrian Bowyer, a Senior Lecturer in mechanical engineering at the University of Bath in the United Kingdom.

The first four official 3D printing machines RepRap project were: "Darwin", released in March 2007, "Mendel", released in October 2009, "Prusa Mendel" and "Huxley" released in 2010, although hundreds of variations exist.[8] The core developers have named each after famous evolutionary biologists, as "the point of RepRap is replication and evolution", however, other variants are often named after individual designers or names they prefer.[9]

23 March 2005 The RepRap blog is started.

Summer 2005 Funding for initial development at the University of Bath is obtained from the UK's Engineering and Physical Sciences Research Council

13 September 2006 The RepRap 0.2 prototype successfully prints the first part of itself, which is subsequently used to replace an identical part originally created by a commercial 3D printer.

9 February 2008 RepRap 1.0 "Darwin" successfully makes at least one instance of over half its total rapid-prototyped parts.

14 April 2008 Possibly the first end-user item is made by a RepRap: a clamp to hold an iPod securely to the dashboard of a Ford Fiesta.

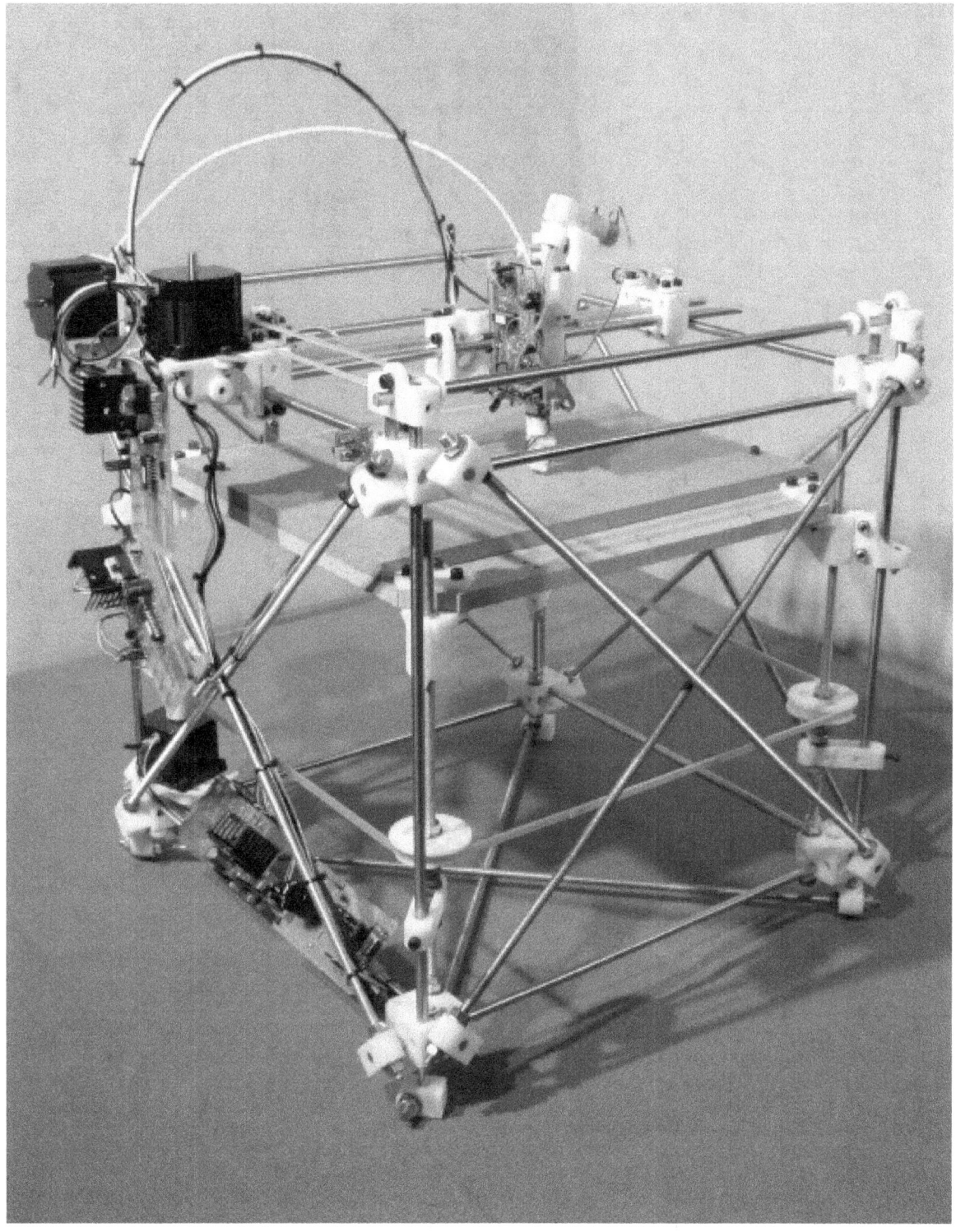

RepRap version 1.0 (Darwin)

29 May 2008 Within a few minutes of being assembled, the first completed "child" machine makes the first part for a "grandchild" at the University of Bath, UK.

23 September 2008 It is reported that at least 100 copies have been produced in various countries. The exact number

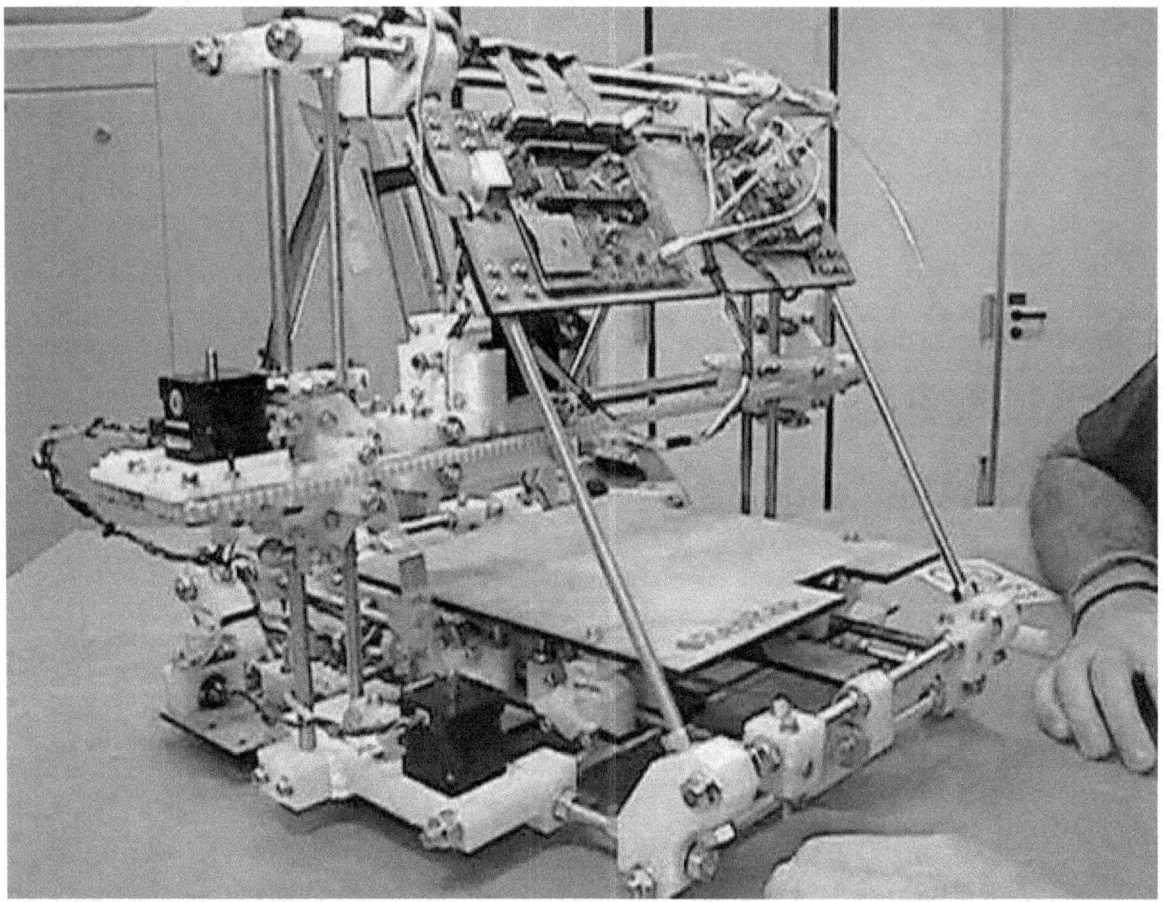

RepRap version 2.0 (Mendel)

of RepRaps in circulation at that time is unknown.[10]

30 November 2008 First documented "in the wild" replication occurs. Replication is completed by Wade Bortz, the first user outside of the developers' team to produce a complete set for another person.

20 April 2009 Announcement of first electronic circuit boards produced automatically with a RepRap, using an automated control system and a swappable head system capable of printing both plastic and conductive solder. Part is later integrated into the RepRap that made it.

2 October 2009 The second generation design, called "Mendel", prints its first part. The Mendel's shape resembles a triangular prism rather than a cube.

13 October 2009 RepRap 2.0 "Mendel" is completed.

27 January 2010 The Foresight Institute announces the "Kartik M. Gada Humanitarian Innovation Prize" for the design and construction of an improved RepRap. There are two prizes, one of US$20,000, and another of $80,000. The administration of the prize is later transferred to Humanity+.[11]

31 August 2010 The third generation design, "Huxley", is officially named. Development is based on a miniaturized version of the Mendel hardware with 30% of the original print volume.

First half 2012 RepRap and RepStrap building and usage are widespread within the tech, gadget, and engineering communities. RepRaps or commercial derivatives have been featured in many mainstream media sources, and are on the permanent watch lists of such tech media as Wired and some influential engineering-professionals' news media.[12]

Adrian Bowyer talking about the RepRap Project at Poptech 2007

Late summer/fall 2012 There has been much focus on smaller startup companies selling derivatives, kits, and assembled systems, and R & D results into new related processes for 3D Printing at orders-of-magnitude-lower prices than current industrial offers. In terms of RepRap research, the most notable result is perhaps the first successful Delta design, Rostock, which is maturing slowly and has an initial working solution for experimentation by self-sourcing builders of some experience. While the Rostock is still in an experimental stage with major revisions almost monthly, it is also near the state of the art, and a radically different design. The latest iterations use OpenBeams, wires (typically Dyneema or Spectra fishing lines) instead of belts, and so forth, which also represents some of the latest trends in RepRaps.

2013 Hobby friendly machines (e.g. recyclebots) have been made to allow hobbyists and small companies to produce their own filament for 3D printing, thus bringing down manufacturing costs and allowing for small-scale recycling and experimenting with different plastics and materials.

56.2   Hardware

As the project was designed by Adrian Bowyer to encourage evolution, many variations have been created.[8][13] As an open source project designers are free to make modifications and substitutions, but they must reshare their improvements. However, RepRap 3D printers generally consist of a thermoplastic extruder mounted on a computer-controlled Cartesian XYZ platform. The platform is built from steel rods and studding connected by printed plastic parts. All three axes are driven by stepper motors, in X and Y via a timing belt and in Z by a leadscrew.

At the heart of the RepRap is the thermoplastic extruder. Early extruders for the RepRap used a geared DC motor driving a screw pressed tightly against plastic filament feedstock, forcing it past a heated melting chamber and through a narrow extrusion nozzle. However, due to their large inertia, DC motors cannot quickly start or stop, and are therefore difficult to control with precision. Therefore, more recent extruders use stepper motors (sometimes geared) to drive the filament, pinching the filament between a splined or knurled shaft and a ball bearing.

RepRap's electronics are based on the popular open-source Arduino platform, with additional boards for controlling stepper motors. The current version electronics use an Arduino-derived Sanguino motherboard, and an additional, customized

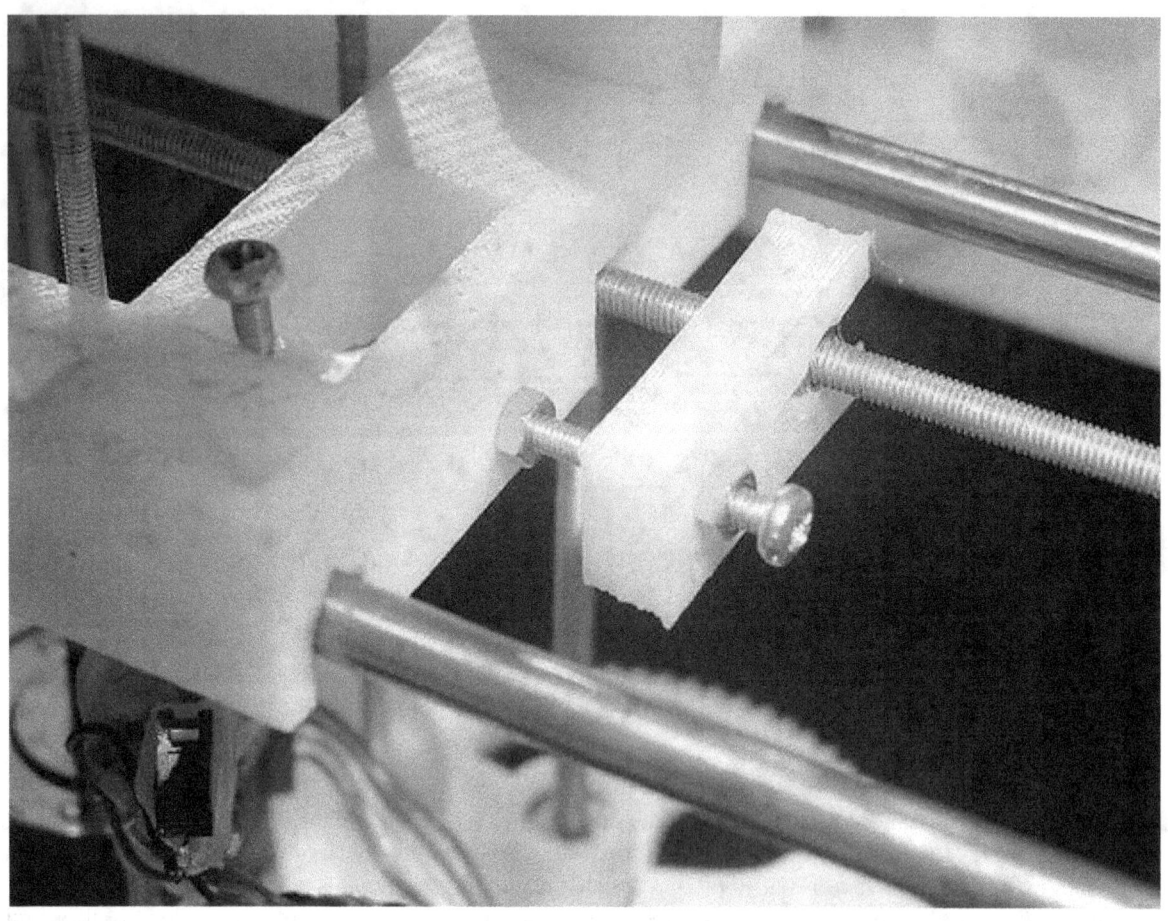

First part ever made by a RepRap to make a RepRap, fabricated by the Zaphod prototype, by Vik Olliver (2006/09/13)

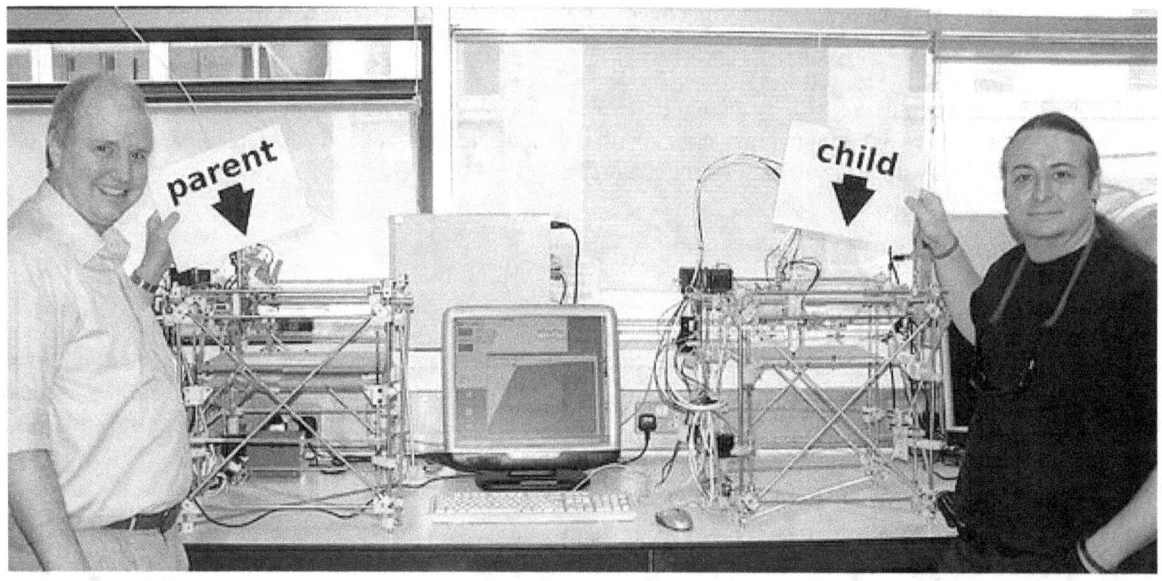

All of the plastic parts for the machine on the right were produced by the machine on the left. Adrian Bowyer (left) and Vik Olliver (right) are members of the RepRap project.

Version 2 'Mendel' holding recently printed physical object next to the driving PC showing a model of the object on-screen

Video of RepRap printing an object

Arduino board for the extruder controller. This architecture allows expansion to additional extruders, each with their own

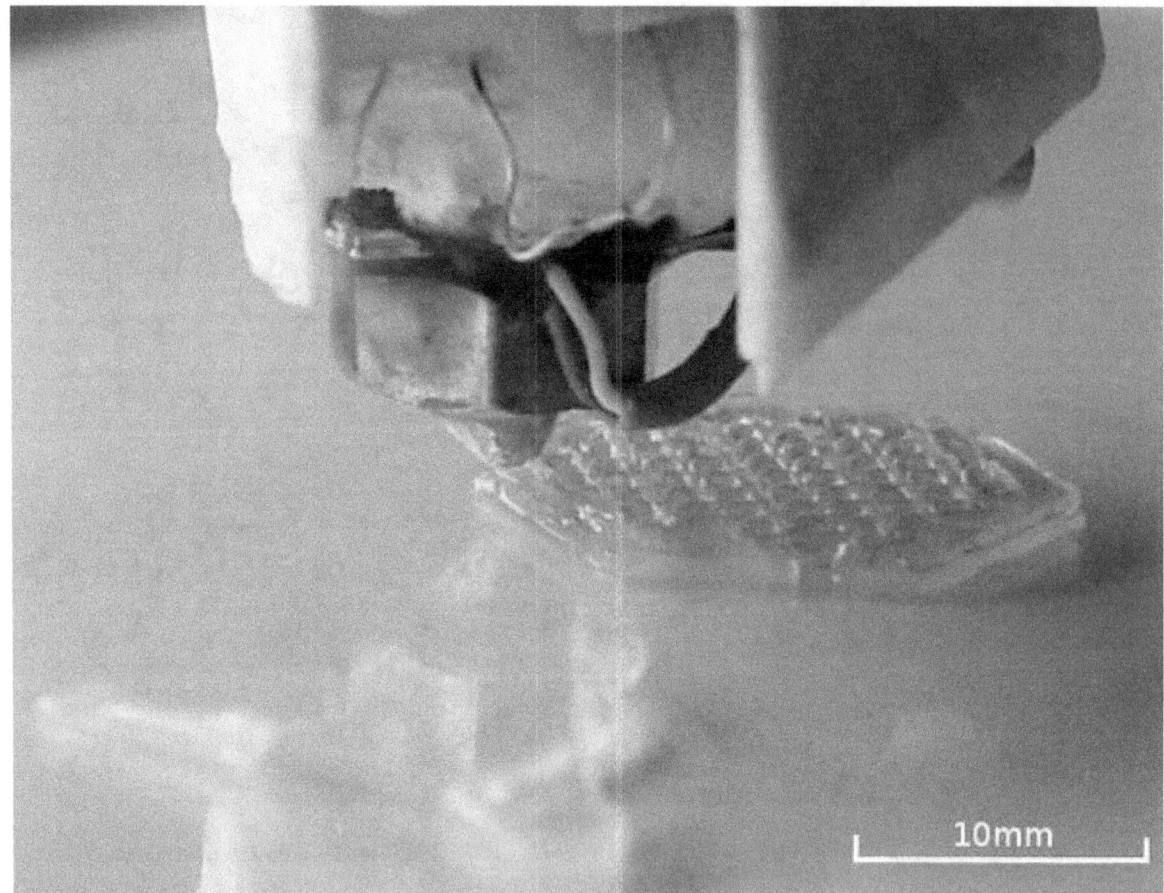

RepRap 0.1 building an object

extruder controller.

56.2.1 Major revisions

The first publicly released RepRap, *Darwin*, has an XY gantry mounted above a moving Z-axis print bed. Darwin's Z axis is constrained by a leadscrew at each corner, all linked together by timing belts to turn in unison. Electronics are mounted on the steel supports of its cuboid exterior, and on a second platform at the base. In an effort to minimize the number of non-printed components (or "vitamins"), Darwin uses printed sliding contact bearings on all of its axes.

Mendel replaced Darwin's sliding bearings with ball bearings, using an exactly constrained design that minimizes friction and tolerates misalignment. It also rearranged the axes, so that the bed slides in the horizontal Y direction, while the extruder moves up and down and in the X direction. This makes Mendel less top-heavy and more compact than Darwin, while also removing the overconstraint of Darwin's four Z axis leadscrews. The build envelope for Mendel is 200 mm (W) × 200 mm (D) × 140 mm (H) or 8" (W) × 8" (D) × 5.5" (H).

One of the more popular RepRap variants from 2013 and beyond is the Rostock delta-style RepRap.[14]

56.3 Software

RepRap has been conceived as a complete replication system rather than simply a piece of hardware. To this end the system includes computer-aided design (CAD) in the form of a 3D modeling system and computer-aided manufacturing

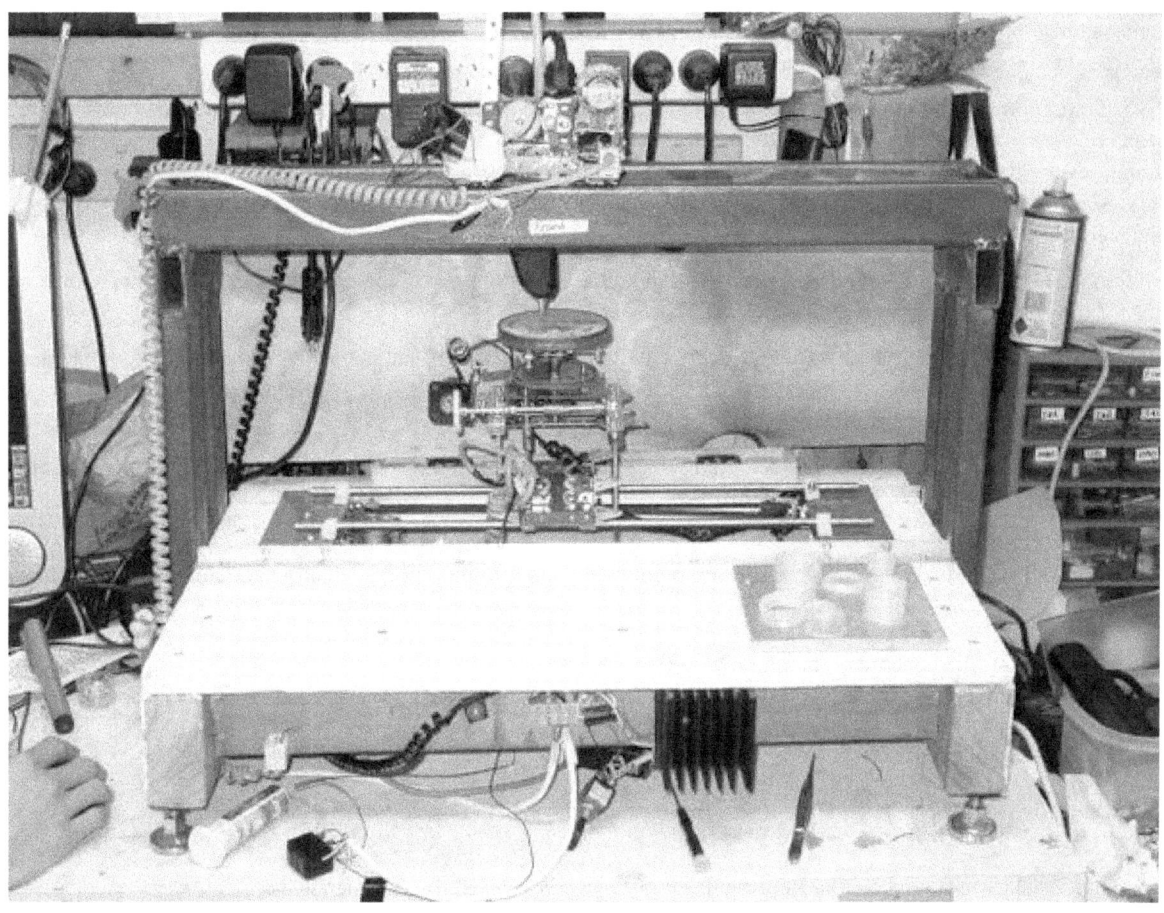

Meccano repstrap of RepRap 0.1 prototype (created by Vik Olliver)

(CAM) software and drivers that convert RepRap users' designs into a set of instructions to the RepRap hardware that turns them into physical objects.

Initially two different CAM toolchains had been developed for the RepRap. The first, simply titled "RepRap Host", was written in Java by lead RepRap developer Adrian Bowyer. The second, "Skeinforge", was written independently by Enrique Perez. Both are complete systems for translating 3D computer models into G-code, the machine language that commands the printer.

Later, other programs like slic3r, pronterface, Cura, repetier host were created. The closed source KISSlicer also seems popular.

Virtually any CAD or 3D modeling program can be used with the RepRap, as long as it is capable of producing STL files.(slic3r also supports .obj and .amf files) Content creators make use of any tools they are familiar with, whether they are commercial CAD programs, such as SolidWorks and Autodesk AutoCAD, Autodesk Inventor, Autodesk 123D Design, Tinkercad, or open-source 3D modeling programs like Blender, OpenSCAD, and FreeCAD.

56.4 Replication materials

RepRaps print objects from ABS, Polylactic acid (PLA), Nylon (possibly not all extruders capable), HDPE, TPE and similar thermoplastics.

Polylactic acid (PLA) has the engineering advantages of high stiffness, minimal warping, and an attractive translucent colour. It is also biodegradable and plant-derived.

The mechanical properties of RepRap printed PLA and ABS have been tested and have been shown to be equivalent to the tensile strengths of proprietary printers.[15]

Unlike in most commercial machines, RepRap users are encouraged to experiment with printing new materials and methods, and to publish their results. Methods for printing novel materials (such as ceramics) have been developed this way. In addition, several RecycleBots have been designed and fabricated to convert waste plastic, such as shampoo containers and milk jugs, into inexpensive RepRap filament.[16] There is some evidence that using this approach of distributed recycling is better for the environment [17][18][19] and be useful for creating "fair trade filament".[20]

In addition, 3D printing products themselves at the point of consumption by the consumer has also been shown to be better for the environment.[21]

The RepRap project has identified polyvinyl alcohol (PVA) as a potentially suitable support material to complement its printing process, although massive overhangs can be made with using thin layers of the primary printing media as support, which are mechanically removed afterwards.

Printing electronics is a major goal of the RepRap project so that it can print its own circuit boards. Several methods have been proposed:

- Wood's metal or Field's metal: low-melting point metal alloys to incorporate electrical circuits into the part as it is being formed.

- Silver/carbon-filled polymers: commonly used for repairs to circuit boards and are being contemplated for use for electrically conductive traces.[22]

- Direct extrusion of solder[23]

- Conductive wires: can be laid into a part from a spool during the printing process

Using a MIG welder as a print head a RepRap deltabot stage can be used to print metals like steel.[24][25]

The RepRap concept can also be applied to a milling machine.[26]

56.5 Construction

Other 3D printer designs (such as the commercial Makerbot) and parts constructed by other means (such as Meccano or wood) may be used to "bootstrap" the RepRap process by building RepRap parts. Many such machines are based on RepRap designs and use RepRap electronics. These are generally known by the name **RepStrap** (for "bootstrap RepRap") by the RepRap community. A RepStrap is any open-hardware rapid-prototyping machine that makes RepRap parts and is itself made by fabrication processes which aren't under the RepRap umbrella yet. Some RepStrap designs are similar to Darwin or Mendel, but they have been modified to be made from laser cut sheets or milled parts. Others, such as the Makerbot, share some design elements with the RepRap (especially electronics), but with a completely reconfigured mechanical structure.

Although the aim of the project is for RepRap to be able to autonomously construct many of its own mechanical components in the near future using fairly low-level resources, several components such as sensors, stepper motors, or microcontrollers are currently non-replicable using the RepRap's 3D printing technology and therefore have to be produced independently of the RepRap self-replicating process. The goal is to asymptotically approach 100% replication over a series of evolutionary generations. As one example, from the onset of the project, the RepRap team has explored a variety of approaches to integrating electrically-conductive media into the product. The future success of this initiative should open the door to the inclusion of connective wiring, printed circuit boards, and possibly even motors in RepRapped products. Variations in the nature of the extruded, electrically-conductive media could produce electrical components with different functions from pure conductive traces, not unlike what was done in the sprayed-circuit process of the 1940s named Electronic Circuit Making Equipment (ECME), described in the article on its designer, John Sargrove. Printed electronics is a related approach. Another non-replicable component is the threaded rods for the linear motions. A current research area is in using replicated Sarrus linkages to replace them.[27]

56.6   Project members

The "Core team" of the project[28] has included:

- Dr. Adrian Bowyer, Former Senior Lecturer, Mechanical Engineering Department, University of Bath

- Michael S. Hart (deceased 2011), creator of Project Gutenberg, Illinois

- Dr. Forrest Higgs, Brosis Innovations, Inc., California

- Rhys Jones, postgraduate, Mechanical Engineering Department, University of Bath

- James Low, undergraduate, Mechanical Engineering Department, University of Bath

- Sebastien Bailard, Ontario

- Simon McAuliffe, New Zealand

- Vik Olliver, Diamond Age Solutions, Ltd., New Zealand

- Ed Sells, postgraduate, Mechanical Engineering Department, University of Bath

- Zach Smith, United States

- Erik de Bruijn, The Netherlands

- Josef Průša, Czech Republic

Project sponsors include:[29]

- Reece Arnott

- The Bath University Innovative Manufacturing Research Centre

- The Engineering and Physical Sciences Research Council

- The Fluorocarbon Co. Ltd

- Michael Ingram

- Lukasz Kaiser

- The Nuffield Foundation

- Carl Witty

56.7   Goals

The stated goal of the RepRap project is to produce a pure self-replicating device not for its own sake, but rather to put in the hands of individuals anywhere on the planet, for a minimal outlay of capital, a desktop manufacturing system that would enable the individual to manufacture many of the artifacts used in everyday life.[1] From a theoretical viewpoint, the project is attempting to prove the hypothesis that "*Rapid prototyping and direct writing technologies are sufficiently versatile to allow them to be used to make a von Neumann Universal Constructor".[30]*

The self-replicating nature of RepRap could also facilitate its viral dissemination and may well facilitate a major paradigm shift in the design and manufacture of consumer products from one of factory production of patented products to one of personal production of un-patented products with open specifications. Opening up product design and manufacturing capabilities to the individual should greatly reduce the cycle time for improvements to products and support a far larger diversity of niche products than the factory production run size can support.

56.8 Education applications

RepRap technology has great potential in educational applications.[31][32][33] RepRaps have already been used for an educational mobile robotics platform.[34] Some authors have claimed that RepRaps offer an unprecedented "revolution" in STEM education.[35] The evidence for such claims comes from both the low cost ability for rapid prototyping in the classroom by students, but also the fabrication of low-cost high-quality scientific equipment from open hardware designs forming open-source labs.[4][5]

56.9 See also

- Additive manufacturing

- 3D printing consumer use

- List of 3D printer manufacturers

- 3D printing services

- Fused Filament Fabrication

- Self-replicating machine

- Disruptive technology

- Distributed manufacturing

- MyMiniFactory

- Open-source appropriate technology

- Open Source Lab (book)

- Fab lab

- MakerBot

- Rapid prototyping

- Recyclebot

- Thingiverse

- Youmagine

- Fab@Home

- G-code

56.10 Notes

[1] Jones, R.; Haufe, P.; Sells, E.; Iravani, P.; Olliver, V.; Palmer, C.; Bowyer, A. (2011). "Reprap-- the replicating rapid prototyper". *Robotica* **29** (1): 177–191. doi:10.1017/s026357471000069x.

[2] http://reprap.org/wiki/RepRapGPLLicence

[3] J. M Pearce, C. Morris Blair, K. J. Laciak, R. Andrews, A. Nosrat and I. Zelenika-Zovko, "3-D Printing of Open Source Appropriate Technologies for Self-Directed Sustainable Development", *Journal of Sustainable Development* **3**(4), pp. 17-29 (2010).

[4] Pearce, Joshua M (2012). "Building Research Equipment with Free, Open-Source Hardware". *Science* **337** (6100): 1303–1304. doi:10.1126/science.1228183.

[5] J.M. Pearce, *Open-Source Lab: How to Build Your Own Hardware and Reduce Research Costs*, Elsevier, 2014.

[6] Sells, E., Smith, Z., Bailard, S., Bowyer, A., & Olliver, V. (2009). Reprap: the replicating rapid prototyper: maximizing customizability by breeding the means of production. Handbook of Research in Mass Customization and Personalization.

[7] Wittbrodt, B.T.; Glover, A.G.; Laureto, J.; Anzalone, G.C.; Oppliger, D.; Irwin, J.L.; Pearce, J.M. (2013). "Life-cycle economic analysis of distributed manufacturing with open-source 3-D printers". *Mechatronics* **23**: 713–726. doi:10.1016/j.mechatronics.2013.06.002.

[8] RepRap Family Tree

[9] RepRap Options, RepRap wiki, http://reprap.org/wiki/RepRap_Options visited 2.26.2014

[10] Matthew Power (2008-09-23). "Mechanical Generation §". Seedmagazine.com. Retrieved 2010-06-04.

[11] "Gada Prizes". humanity+. Retrieved 25 April 2011.

[12] "Ingeniøren". Ingeniøren media. 2012-09-26. Retrieved 2012-09-26.

[13] Chulilla, J. L. (2011). "The Cambrian Explosion of Popular 3D Printing". *International Journal of Interactive Multimedia and Artificial Intelligence* **1**: 4.

[14] http://reprap.org/wiki/Rostock

[15] B.M. Tymrak, M. Kreiger, J. M. Pearce "Mechanical properties of components fabricated with open-source 3-D printers under realistic environmental conditions" *Materials & Design*, 58, pp. 242-246 (2014). doi 10.1016/j.matdes.2014.02.038. open access

[16] Baechler, Christian; DeVuono, Matthew; Pearce, Joshua M. "Distributed Recycling of Waste Polymer into RepRap Feedstock". *Rapid Prototyping Journal* **19** (2): 118–125.

[17] Kreiger, M., Anzalone, G. C., Mulder, M. L., Glover, A., & Pearce, J. M. (2013). Distributed Recycling of Post-Consumer Plastic Waste in Rural Areas. MRS Online Proceedings Library, 1492, mrsf12-1492. open access

[18] The importance of the Lyman Extruder, Filamaker, Recyclebot and Filabot to 3D printing – VoxelFab, 2013.

[19] M. Kreiger, G. C. Anzalone, M. L. Mulder, A. Glover and J. M Pearce (2013). Distributed Recycling of Post-Consumer Plastic Waste in Rural Areas. MRS Online Proceedings Library, 1492, mrsf12-1492-g04-06 doi:10.1557/opl.2013.258. open access

[20] Feeley, S. R.; Wijnen, B.; Pearce, J. M. (2014). "Evaluation of Potential Fair Trade Standards for an Ethical 3-D Printing Filament". *Journal of Sustainable Development* **7** (5): 1–12. doi:10.5539/jsd.v7n5p1.

[21] M. Kreiger and J.M. Pearce. (2013). Environmental Life Cycle Analysis of Distributed 3-D Printing and Conventional Manufacturing of Polymer Products, *ACS Sustainable Chemistry & Engineering* **1** (12), (2013) pp. 1511–1519. DOI: 10.1021/sc400093k 10.1021/sc400093k Open access

[22] Simon J. Leigh, Robert J. Bradley, Christopher P. Purssell, Duncan R. Billson, David A. Hutchins A Simple, Low-Cost Conductive Composite Material for 3D Printing of Electronic Sensors

[23] RepRap blog 2009 visited 2/26/2014

[24] An Inexpensive Way to Print Out Metal Parts - The New York Times

[25] Gerald C. Anzalone, Chenlong Zhang, Bas Wijnen, Paul G. Sanders and Joshua M. Pearce, "Low-Cost Open-Source 3-D Metal Printing" *IEEE Access*, 1, pp.803-810, (2013). doi: 10.1109/ACCESS.2013.2293018 open access preprint

[26] Kostakis, V., & Papachristou, M. (2013). Commons-based peer production and digital fabrication: The case of a RepRap-based, Lego-built 3D printing-milling machine. Telematics and Informatics.

[27] "I, replicator". *New Scientist*. 29 May 2010.

[28] "The Core Team - who we are", reprap.org/wiki

[29] "Acknowledgements" reprap.org/wiki

[30] "RepRap—the Replication Rapid Prototyper Project, IdMRC" (PDF). Retrieved 2007-02-19.

[31] Chelsea Schelly,Gerald Anzalone,Bas Wijnen,Joshua M. Pearce, (2015). Open-source 3-D printing Technologies for education: Bringing Additive Manufacturing to the Classroom.*Journal of Visual Languages & Computing.* 2015; 28226–237. open access

[32] Grujović, N., Radović, M., Kanjevac, V., Borota, J., Grujović, G., & Divac, D. (2011, September). 3D printing technology in education environment. *In 34th International Conference on Production Engineering* (pp. 29-30).

[33] Mercuri, R., & Meredith, K. (2014, March). An educational venture into 3D Printing. In Integrated STEM Education Conference (ISEC), 2014 IEEE (pp. 1-6). IEEE.

[34] Gonzalez-Gomez, J., Valero-Gomez, A., Prieto-Moreno, A., & Abderrahim, M. (2012). A new open source 3d-printable mobile robotic platform for education. In *Advances in autonomous mini robots* (pp. 49-62). Springer Berlin Heidelberg.

[35] J. Irwin, J.M. Pearce, D. Opplinger, and G. Anzalone. The RepRap 3-D Printer Revolution in STEM Education,*121st ASEE Annual Conference and Exposition, Indianapolis, IN.* Paper ID #8696 (2014).

56.11 References

- Replication revolutionary. *New Electronics*, 12 December 2006.
- 3D printer to churn out copies of itself. Celeste Biever, *New Scientist*, 18 March 2005
- The machine that can copy anything. Simon Hooper, CNN, 2 June 2005
- Self Replicating Robots And The Developing World. KnowProSE.com, 5 June 2005
- Interview with Vik Olliver about RepRap, September 2006
- Chinese Growth Hurdles toward a New Great Wall
- Canadian Broadcasting Corporation audio interview with Adrian Bowyer

56.12 External links

- Official website
- Video of a talk by Adrian Bowyer on RepRap

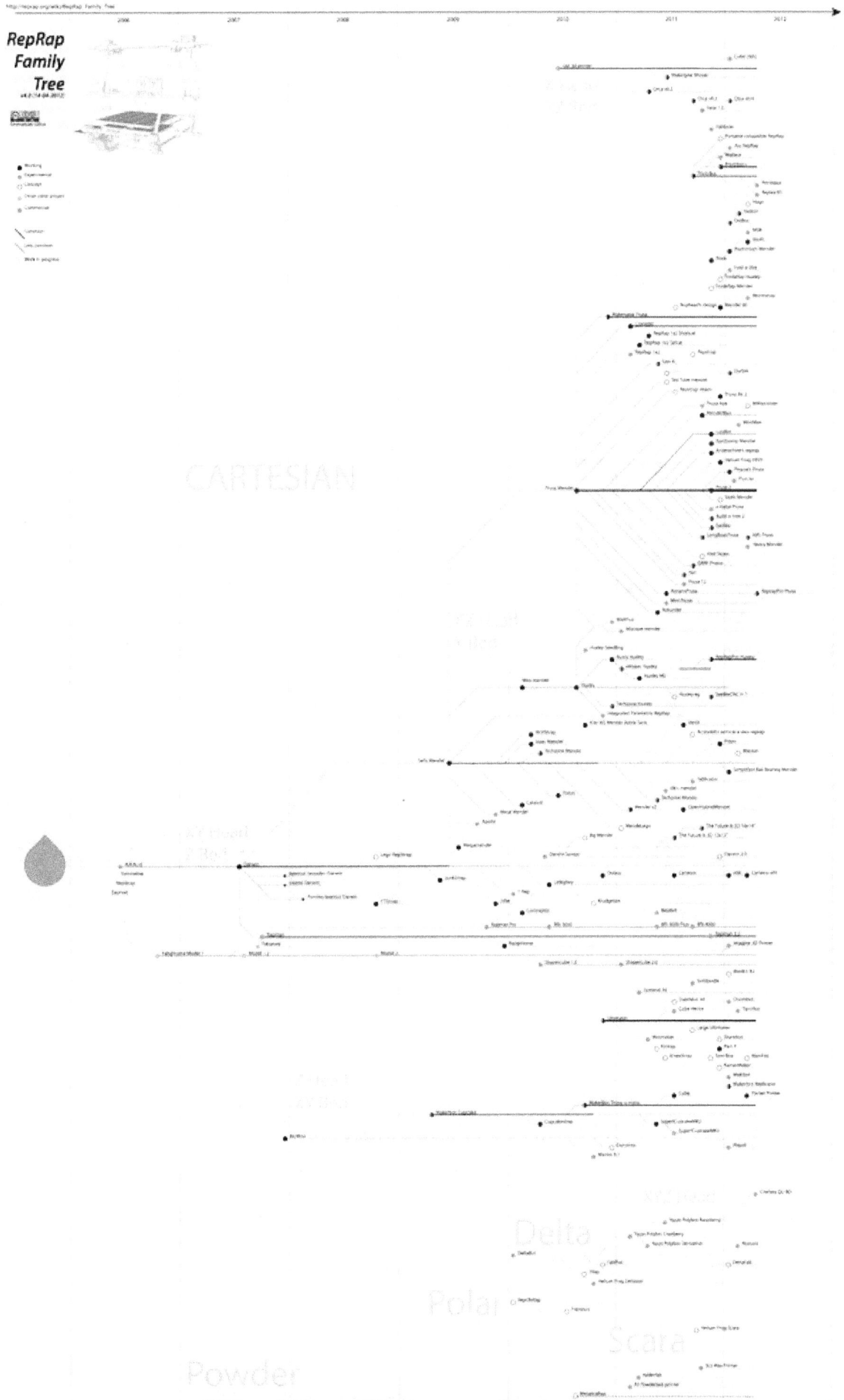

Chapter 57

Robocasting

Robocasting a set of simple bars

Robocasting or **Direct Ink Writing** (**DIW**) is an additive manufacturing technique in which a filament of 'ink' is extruded from a nozzle, forming an object layer by layer. The technique was first developed in the United States in 1996 as a method to allow geometrically complex ceramic green bodies to be produced by additive manufacturing.[1] In robocasting, a 3D CAD model is divided up into layers in a similar manner to other additive manufacturing techniques. A fluid (typically a ceramic slurry), referred to as an 'ink', is then extruded through a small nozzle as the nozzle's position is controlled, drawing out the shape of each layer of the CAD model. The ink exits the nozzle in a liquid-like state but retains its shape immediately, exploiting the rheological property of shear thinning. It is distinct from fused deposition modelling as it does not rely on the solidification or drying to retain its shape after extrusion.

57.1 Process

Robocasting begins with a software process which slices an STL file (stereolithography file format) into layers of similar thickness to the nozzle diameter. The part is produced by extruding a continuous filament of ink material in the shape required to fill the first layer. Next, either the stage is moved down or the nozzle is moved up and the next layer is deposited in the required pattern. This is repeated until the 3d part is complete. Numerically controlled mechanisms are typically used to move the nozzle in a calculated tool-path generated by a computer-aided manufacturing (CAM) software package. Stepper motors or servo motors are usually employed to move the nozzle with precision as fine as nanometers.[2]

The part is typically very fragile and soft at this point. Drying, debinding and sintering usually follow to give the part the desired mechanical properties.

Depending on the ink composition, printing speed and printing environment, robocasting can typically deal with moderate overhangs and large spanning regions many times the filament diameter in length, where the structure is unsupported from below.[3] This allows intricate periodic 3D scaffolds to be printed with ease, a capability which is not possessed by other additive manufacturing techniques. These parts have shown extensive promise in fields of photonic crystals, bone transplants, catalyst supports and filters. Furthermore supporting structures can also be printed from a "fugitive ink" which is easily removed. This allows almost any shape to be printed in any orientation.

57.2 Applications

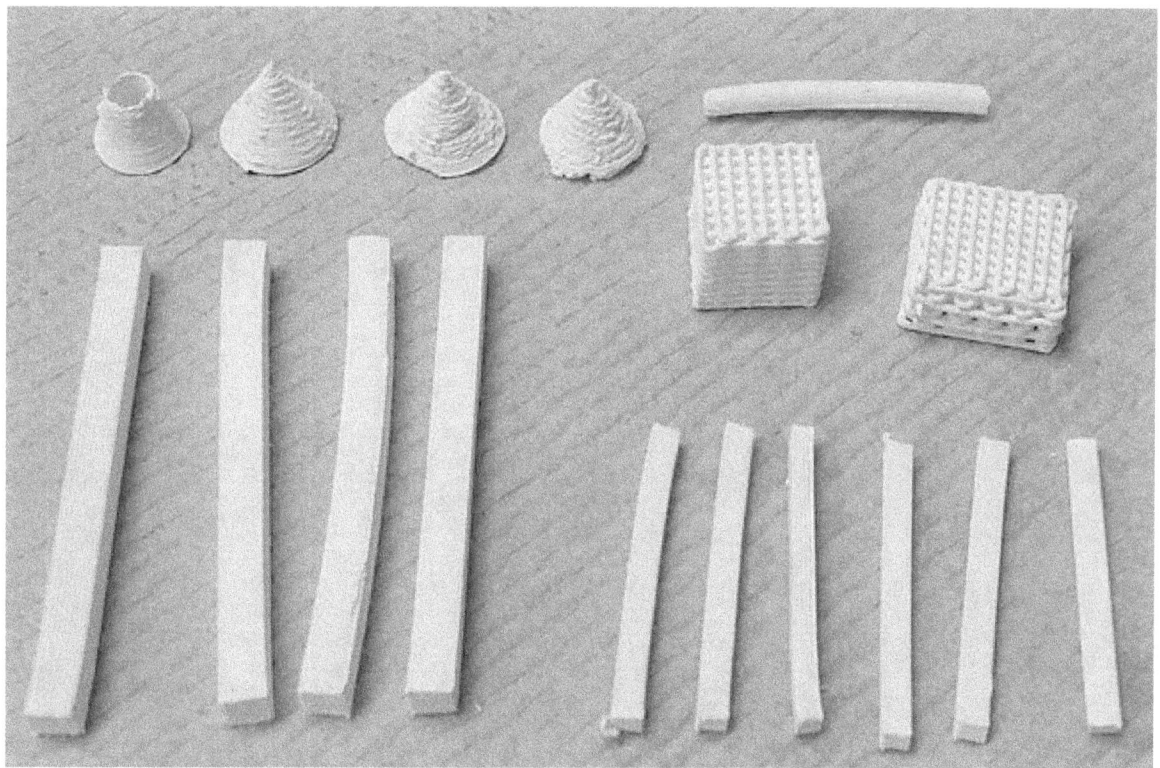

An array of simple alumina geometries created by robocasting.

The technique can produce non-dense ceramic bodies which are very fragile and must be sintered before they can be used for most applications, analogous to a wet clay ceramic pot before being fired. A wide variety of different geometries can be formed from the technique, from solid bars of material to intricate microscale "scaffolds".[4] To date the most researched application for robocasting is in the production of biologically compatible tissue implants. "Woodpile" stacked lattice structures can be formed quite easily which allow bone and other tissues in the human body to grow and eventually replace the transplant. With various medical scanning techniques the precise shape of the missing tissue was be established and input into 3d modelling software and printed. Calcium phosphate glasses and hydroxyapetite have been extensively explored as candidate materials due to their biocompatibility and structural similarity to bone.[5] Other potential applications include the production of specific high surface area structures, such as catalyst beds or fuel cell electrolytes.[6] Advanced metal matrix- and ceramic matrix- load bearing composites can be formed by infiltrating woodpile bodies with molten glasses, alloys or slurries.

Robocasting has also been used to deposit polymer and sol-gel inks through much finer nozzle diameters (<2μm) than is possible with ceramic inks.[2]

57.3 References

[1] Stuecker, J (2004). "Advanced Support Structures for Enhanced Catalytic Activity". *Industrial & Engineering Chemistry Research* **43** (1): 51–55. doi:10.1021/ie030291v.

[2] Xu, Mingjie; Gratson, Gregory M.; Duoss, Eric B.; Shepherd, Robert F.; Lewis, Jennifer A. (2006). "Biomimetic silicification of 3D polyamine-rich scaffolds assembled by direct ink writing". *Soft Matter* **2** (3): 205. doi:10.1039/b517278k. ISSN 1744-683X.

[3] Smay, James E.; Cesarano, Joseph; Lewis, Jennifer A. (2002). "Colloidal Inks for Directed Assembly of 3-D Periodic Structures". *Langmuir* **18** (14): 5429–5437. doi:10.1021/la0257135. ISSN 0743-7463.

[4] Lewis, Jennifer (2006). "Direct Ink Writing of 3D Functional Materials". *Advanced Functional Materials* **16** (17): 2193–2204. doi:10.1002/adfm.200600434.

[5] Miranda, P (2008). "Mechanical properties of calcium phosphate scaffolds fabricated by robocasting.". *Journal of Biomedical Materials* **85** (1): 218–227. doi:10.1002/jbm.a.31587. PMID 17688280.

[6] Kuhn, M.; Napporn, T.; Meunier, M.; Vengallatore, S.; Therriault, D. (2008). "Direct-write microfabrication of single-chamber micro solid oxide fuel cells". *Journal of Micromechanics and Microengineering* **18**: 015005. doi:10.1088/0960-1317/18/1/015005.

57.4 External links

- Robocasting, MIT Technology Review

Chapter 58

Sanguino3 G-Code

Sanguino3 G-Code is the protocol by which 3rd-generation RepRap Project electronics communicate with their host machine, as well as the protocol by which the RepRap host communicates with its subsystems.[1] It can also be written in a binary format to storage for later replay, usually in a file with a ".s3g" extension.[2]

The protocol is intended as a simplification of G-code, to ease processing by the somewhat limited CPU of the Sanguino, an Arduino-based controller.

[1] RepRap Project, RepRap Generation 3 Protocol Specification v1, retrieved 2011-05-01, no longer works, dead follow link 2013-02-17

[2] ReplicatorG, Sangiuno3G, retrieved 2011-05-01

Chapter 59

Sculpteo

Sculpteo is a French company specialized in 3D printing in the cloud.[1] Sculpteo offers an online 3D printing service, using rapid prototyping and a manufacturing process [2] involving laser sintering or stereo lithography. The company was founded in June 2009 by Eric Carreel co-founder of Inventel,[3] acquired by Technicolor in 2005 [4] and Withings, Clement Moreau [5] and Jacques Lewiner.

59.1   Purpose

Sculpteo offers a 3D printing service which is accessible online.[6] Starting with a 3D file, Sculpteo creates real objects such as interior decorations, characters, robots, miniatures and models.[7]

59.2   History and background

- In January 2014 at Las Vegas Consumer Electronics Show Sculpteo unveils 3D Batch Control, a cloud 3D printing service especially designed for professionals and businesses in need of short-run manufacturing. It allows users to upload a 3D file, change the size and dimensions of the object directly within the browser, select a printing material, and order their design to be 3D printed and shipped.[8]

- In January 2012, Sculpteo launched the 3D printing Cloud Engine and the Sculpteo app at the Consumer Electronics Show in Las Vegas. The app transforms human data into a 3D printed object using an iOS device.[9]

- In October 2011 Sculpteo launched a 3D silver printing service using Sculpteo's silver process™

- In June 2011, Sculpteo and 3DVIA announced that they had established a direct printing service provided by Sculpteo via the 3Dvia portal.[10][11]

- In March 2011, Sculpteo launched Pro.Sculpteo an online 3D printing service for professionals.[12]

- In January 2011, Sculpteo launched their online 3D printing service to the public:[13] the ability to make your own avatar. Starting with simple 2D photos which are then modeled in 3D.[14]

- The company was created in June 2009.[15]

59.3   Awards

• In January 2013, Sculpteo was honoured by the Consumer Electronics Show with a Best of CES Innovations Award for 3DPCase, its new 3D printing mobile application[16] 3DPcase.[17][18]

• In 2012, Sculpteo was honoured by Observeur du Design label from Observeur du Design [19] which can demonstrate how the design world can be seen to recognise the work which Sculpteo does.

59.4   References

[1] Adam Ludwig (2012-04-18). "Sculpteo Takes 3D Printing to the Cloud". Forbes. Retrieved 2012-07-31.

[2] Todd Wasserman 91 (2011-04-10). "Will 3D Printing Reboot Manufacturing? [PICS]". Mashable.com. Retrieved 2012-07-31.

[3] Les Echos. "Les Echos - Ses décodeurs ont conquis Thomson - Archives". Archives.lesechos.fr. Retrieved 2012-07-31.

[4] Les Echos. "Les Echos - Thomson acquiert la start-up française Inventel - Archives". Archives.lesechos.fr. Retrieved 2012-07-31.

[5] Les Echos. "Les Echos - Sculpteo réalise vos objets à partir de fichiers 3D - Archives". Lesechos.fr. Retrieved 2012-07-31.

[6]

[7] Leach, Anna (2011-03-17). "Order a 3 inch model of yourself from the internet: Sculpteo Mini Me is definitely worth it : Shiny Shiny". Shinyshiny.tv. Retrieved 2012-07-31.

[8] http://www.engadget.com/2014/01/05/sculpteo-3d-printing-batch-control/

[9] "APP OF THE DAY: Scultpeo review (iOS)". Pocket-lint. 2012-01-12. Retrieved 2012-07-31.

[10] Sterling, Bruce (2011-06-27). "Spime Watch: Dassault Systèmes' 3DVIA and Sculpteo | Beyond The Beyond". Wired.com. Retrieved 2012-07-31.

[11] par Julien Mechin 27 juin 2011 (2011-06-27). "Sculpteo signe un accord historique avec Dassault Systèmes". Fr.techcrunch.com. Retrieved 2012-07-31.

[12] "L'impression 3D à portée de tous ?". Lemonde.fr. Retrieved 2012-07-31.

[13] "Les robots sculpteurs débarquent avec des prix défiant tout concurrence". Latribune.fr. 2011-08-18. Retrieved 2012-07-31.

[14] "Your own mini-me? Sculpteo lets you print a real-life 3D avatar". Eu.techcrunch.com. Retrieved 2013-10-29.

[15] lefigaro.fr. "Le Figaro - Sciences et Technologies : Une imprimante sculpte vos objets en 3D". Lefigaro.fr. Retrieved 2012-07-31.

[16] "2013 - 2014 International CES, January 7 - 10". Cesweb.org. Retrieved 2013-10-29.

[17] "Les start-up françaises encore à l'honneur au CES de Las Vegas". Lesechos.fr. 2013-01-07. Retrieved 2013-10-29.

[18] "Sculpteo awarded Best of the Best at CES 2013 Las Vegas - Personalize | 3D Printing News, 3D Printer reviews & Additive Manufacturing Interviews". Prsnlz.me. 2012-11-13. Retrieved 2013-10-29.

[19]

59.5   See also

- 3D Printing Marketplace

- Pinshape

- 3DLT

- Sketchfab

- Shapeways

- Thingiverse

- Materialise NV

- Kraftwurx

Chapter 60

Selective heat sintering

Selective heat sintering (**SHS**) is a type of additive manufacturing process. It works by using a thermal printhead to apply heat to layers of powdered thermoplastic. When a layer is finished, the powder bed moves down, and an automated roller adds a new layer of material which is sintered to form the next cross-section of the model. SHS is best for manufacturing inexpensive prototypes for concept evaluation, fit/form and functional testing. SHS is a Plastics additive manufacturing technique similar to selective laser sintering (SLS), the main difference being that SHS employs a less intense thermal printhead instead of a laser, thereby making it a cheaper solution, and able to be scaled down to desktop sizes.[1]

60.1 References

[1] "How Selective Heat Sintering Works". THRE3D.com. Retrieved 3 February 2014.

Chapter 61

Selective laser melting

Selective laser melting is an additive manufacturing process that uses 3D CAD data as a digital information source and energy in the form of a high-power laser beam, to create three-dimensional metal parts by fusing fine metal powders together. Manufacturing applications in aerospace or medical orthopedics are being pioneered.

61.1   History

Selective laser melting started in 1995 at the Fraunhofer Institute ILT in Aachen, Germany, with a German research project, resulting in the so-called basic ILT SLM patent DE 19649865. Already during its pioneering phase Dr. Dieter Schwarze and Dr. Matthias Fockele from F&S Stereolithographietechnik GmbH located in Paderborn collaborated with the ILT researchers Dr. Wilhelm Meiners and Dr. Konrad Wissenbach. In the early 2000s F&S entered into a commercial partnership with MCP HEK GmbH (later on named MTT Technology GmbH and then SLM Solutions GmbH) located in Luebeck in northern Germany. Today Dr. Dieter Schwarze is with SLM Solutions GmbH and Dr. Matthias Fockele founded Realizer GmbH.

The ASTM International F42 standards committee has grouped selective laser melting into the category of "laser sintering", although this is an acknowledged misnomer because the process fully melts the metal into a solid homogeneous mass, unlike selective laser sintering (SLS) and direct metal laser sintering (DMLS), which are true sintering processes. A similar process is electron beam melting (EBM), which uses an electron beam as energy source.

61.2   Process

The process starts by slicing the 3D CAD file data into layers, usually from 20 to 100 micrometres thick, creating a 2D image of each layer; this file format is the industry standard .stl file used on most layer-based 3D printing or stereolithography technologies. This file is then loaded into a file preparation software package that assigns parameters, values and physical supports that allow the file to be interpreted and built by different types of additive manufacturing machines.

With selective laser melting, selectively melts thin layers of atomized fine metal powder are evenly distributed using a coating mechanism onto a substrate plate, usually metal, that is fastened to an indexing table that moves in the vertical (Z) axis. This takes place inside a chamber containing a tightly controlled atmosphere of inert gas, either argon or nitrogen at oxygen levels below 500 parts per million. Once each layer has been distributed, each 2D slice of the part geometry is fused by selectively melting the powder. This is accomplished with a high-power laser beam, usually an ytterbium fiber laser with hundreds of watts. The laser beam is directed in the X and Y directions with two high frequency scanning mirrors. The laser energy is intense enough to permit full melting (welding) of the particles to form solid metal. The process is repeated layer after layer until the part is complete.

61.3 Materials

Most machines operate with a build chamber of 250 mm in X & Y and up to 350 mm Z, although larger machines up to 500 mm X,Y,Z and smaller machines do exist. The types of materials that can be processed include stainless steel, tool steel, cobalt chrome, titanium and aluminium. All must exist in atomized form and exhibit certain flow characteristics in order to be process capable.

61.4 Applications

The types of applications most suited to the selective laser melting process are complex geometries & structures with thin walls and hidden voids or channels on the one hand or low lot sizes on the other hand. Advantage can be gained when producing hybrid forms where solid and partially formed or lattice type geometries can be produced together to create a single object, such as a hip stem or acetabular cup or other orthopedic implant where oseointegration is enhanced by the surface geometry. Much of the pioneering work with selective laser melting technologies is on lightweight parts for aerospace[1] where traditional manufacturing constraints, such as tooling and physical access to surfaces for machining, restrict the design of components. SLM allows parts to be built additively to form near net shape components rather than by removing waste material.

Traditional manufacturing techniques have a relatively high set-up cost (e.g. for creating a mold). While SLM has a high cost per part (mostly because it is time-intensive), it is advisable if only very few parts are to be produced. This is the case e.g. for spare parts of old machines (like vintage cars) or individual products like implants.

Tests by NASA's Marshall Space Flight Center, which is experimenting with the technique to make some difficult-to-fabricate parts from nickel alloys for the J-2X and RS-25 rocket engines, show that difficult to make parts made with the technique are somewhat weaker than forged and milled parts but often avoid the need for welds which are weak points.[1]

61.5 Potential

Selective laser melting or additive manufacturing, sometimes referred to as rapid manufacturing or rapid prototyping, is in its infancy with relatively few users in comparison to conventional methods such as machining, casting or forging metals, although those that are using the technology have become highly proficient. Like any process or method selective laser melting must be suited to the task at hand. Markets such as aerospace or medical orthopedics have been evaluating the technology as a manufacturing process. Barriers to acceptance are high and compliance issues result in long periods of certification and qualification. This is demonstrated by the lack of fully formed international standards by which to measure the performance of competing systems. The standard in question is ASTM F2792-10 Standard Terminology for Additive Manufacturing Technologies.

61.6 See also

- List of notable 3D printed weapons and parts

- 3D printing

- Desktop manufacturing

- Digital fabricator

- Direct digital manufacturing

- Rapid manufacturing

- Selective laser sintering

- Solid freeform fabrication

- Stereolithography

61.7 References

[1] Larry Greenemeier (November 9, 2012). "NASA Plans for 3-D Printing Rocket Engine Parts Could Boost Larger Manufacturing Trend". *Scientific American*. Retrieved November 13, 2012.

- ASTM F2792-10 Standard Terminology for Additive Manufacturing Technologies

- Abe, F., Costa Santos, E., Kitamura, Y., Osakada, K., Shiomi, M. 2003. Influence of forming conditions on the titanium model in rapid prototyping with the selective laser melting process. Proceedings of the Institution of Mechanical Engineers, Part C: Journal of Mechanical Engineering Science 217 (1), pp. 119–126.

- Gibson, I. Rosen, D.W. and Stucker, B. (2010) Additive Manufacturing Technologies: Rapid Prototypingto Direct Digital Manufacturing. New York, Hiedelberg, Dordrecht, London: Springer. ISBN

- Wohlers, T. Wohlers Report 2010: Additive Manufacturing State of the Industry: Annual World Wide Progress Report. Fort Collins: Wohlers Associates.

61.8 Further reading

- "How Selective Laser Melting Works". THRE3D.com. Retrieved 11 February 2014.

Chapter 62

Selective laser sintering

An SLS machine being used at the Centro Renato Archer in Brazil.

Selective Laser Sintering (**SLS**) is an additive manufacturing (AM) technique that uses a laser as the power source to sinter powdered material (typically metal), aiming the laser automatically at points in space defined by a 3D model, binding the material together to create a solid structure. It is similar to direct metal laser sintering (DMLS); the two are instantiations of the same concept but differ in technical details. Selective laser melting (SLM) uses a comparable concept, but in SLM the material is fully melted rather than sintered,[1] allowing different properties (crystal structure, porosity, and so on). SLS (as well as the other mentioned AM techniques) is a relatively new technology that so far has mainly been used for rapid prototyping and for low-volume production of component parts. Production roles are expanding as the commercialization of AM technology improves.

62.1 History

Selective laser sintering (SLS) was developed and patented by Dr. Carl Deckard and academic adviser, Dr. Joe Beaman at the University of Texas at Austin in the mid-1980s, under sponsorship of DARPA.[2] Deckard and Beaman were involved in the resulting start up company DTM, established to design and build the SLS machines. In 2001, 3D Systems the biggest competitor of DTM and SLS technology acquired DTM.[3] The most recent patent regarding Deckard's SLS technology was issued 28 January 1997 and expired 28 Jan 2014.[4]

A similar process was patented without being commercialized by R. F. Housholder in 1979.[5]

62.2 Technology

An additive manufacturing layer technology, SLS involves the use of a high power laser (for example, a carbon dioxide laser) to fuse small particles of plastic, metal, ceramic, or glass powders into a mass that has a desired three-dimensional shape. The laser selectively fuses powdered material by scanning cross-sections generated from a 3-D digital description of the part (for example from a CAD file or scan data) on the surface of a powder bed. After each cross-section is scanned, the powder bed is lowered by one layer thickness, a new layer of material is applied on top, and the process is repeated until the part is completed.

Because finished part density depends on peak laser power, rather than laser duration, a SLS machine typically uses a pulsed laser. The SLS machine preheats the bulk powder material in the powder bed somewhat below its melting point, to make it easier for the laser to raise the temperature of the selected regions the rest of the way to the melting point.[6]

In contrast with some other additive manufacturing processes, such as stereolithography (SLA) and fused deposition modeling (FDM), which most often require special support structures to fabricate overhanging designs, SLS does not need a separate feeder for support material because the part being constructed is surrounded by unsintered powder at all times, this allows for the construction of previously impossible geometries. Also, since the machines chamber is always filled with powder materialthe fabrication of multiple parts has a far lower impact on the overall difficulty and price of the design because through a technique known as 'Nesting' multiple parts can be positioned to fit within the boundaries of the machine. One design aspect which should be observed however is that with SLS it is 'impossible' to fabricate a hollow but fully enclosed element. This is because the unsintered powder within the element can't be drained.

Since patents have started to expire, affordable home printers have become possible, but the heating process is still an obstacle, with a power consumption of up to 5 kW and temperatures having to be controlled within 2 °C for the three stages of preheating, melting and storing before removal.

62.3 Materials and applications

Some SLS machines use single-component powder, such as direct metal laser sintering. Powders are commonly produced by ball milling. However, most SLS machines use two-component powders, typically either coated powder or a powder mixture. In single-component powders, the laser melts only the outer surface of the particles (surface melting), fusing the solid non-melted cores to each other and to the previous layer.[6]

Compared with other methods of additive manufacturing, SLS can produce parts from a relatively wide range of commercially available powder materials. These include polymers such as nylon (neat, glass-filled, or with other fillers) or polystyrene, metals including steel, titanium, alloy mixtures, and composites and green sand. The physical process can be full melting, partial melting, or liquid-phase sintering. Depending on the material, up to 100% density can be achieved with material properties comparable to those from conventional manufacturing methods. In many cases large numbers of parts can be packed within the powder bed, allowing very high productivity.

SLS technology is in wide use around the world due to its ability to easily make very complex geometries directly from digital CAD data. While it began as a way to build prototype parts early in the design cycle, it is increasingly being used in limited-run manufacturing to produce end-use parts. One less expected and rapidly growing application of SLS is its use in art.

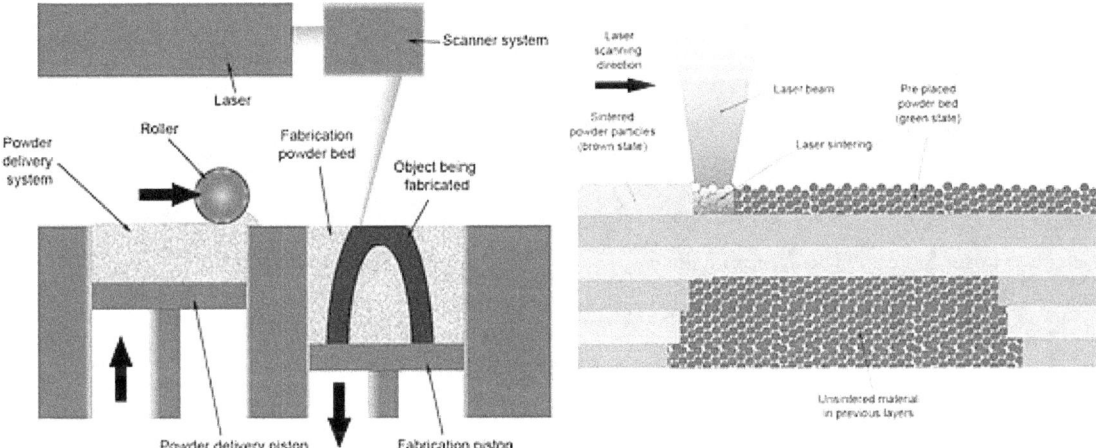

Selective laser sintering process

62.4   See also

- 3D printing

- Desktop manufacturing

- Digital fabricator

- Direct digital manufacturing

- Fab lab

- Instant manufacturing, also known as "direct manufacturing" or "on-demand manufacturing"

- Rapid manufacturing

- Rapid prototyping

- RepRap Project

- Solid freeform fabrication

- Von Neumann universal constructor

62.5   References

[1] "How Selective Laser Sintering Works". THRE3D.com. Retrieved 7 February 2014.

[2] Deckard, C., "Method and apparatus for producing parts by selective sintering", U.S. Patent 4,863,538, filed October 17, 1986, published September 5, 1989.

[3] Lou, Alex and Grosvenor, Carol "Selective Laser Sintering, Birth of an Industry", *The University of Texas*, December 07, 2012. Retrieved on March 22, 2013.

[4] US5597589

[5] Housholder, R., "Molding Process", U.S. Patent 4,247,508, filed December 3, 1979, published January 27, 1981.

[6] Prasad K. D. V. Yarlagadda; S. Narayanan (February 2005). *GCMM 2004: 1st International Conference on Manufacturing and Management*. Alpha Science Int'l. pp. 73–. ISBN 978-81-7319-677-5. Retrieved 18 June 2011.

62.6 External links

- DMLS – DEVELOPMENT HISTORY AND STATE OF THE ART

- Selective Laser Sintering, Birth of an Industry

Chapter 63

Solid Ground Curing

Solid ground curing (SGC) is a photo-polymer-based additive manufacturing (or **3D printing**)[1] technology used for producing models, prototypes, patterns, and production parts, in which the production of the layer geometry is carried out by means of a high-powered UV lamp through a mask. As the basis of solid ground curing is the exposure of each layer of the model by means of a lamp through a mask, the processing time for the generation of a layer is independent of the complexity of the layer.[2] SGC was developed and commercialized by Cubital Ltd. of Israel in 1986[3] in the alternative name of **Solider System**. While the method offered good accuracy and a very high fabrication rate, it suffered from high acquisition and operating costs due to system complexity. This led to poor market acceptance. While the company still exists, systems are no longer being sold. Nevertheless, it's still an interesting example of the many technologies other than stereolithography, its predeceasing rapid prototyping process that also utilizes photo-polymer materials.[4] Though Objet Geometries Ltd. of Israel retains intellectual property of the process after the closure of Cubital Ltd. in 2002,[5] the technology is no longer being produced.

63.1 Technology

Solid ground curing utilizes the general process of hardening of photopolymers by a complete lighting and hardening of the entire surface, using specially prepared masks.[6] In SGC process, each layer of the prototype is cured by exposing to an ultra violet (UV) lamp instead of by laser scanning. So that, every portion in a layer are simultaneously cured and do not require any post-curing processes. The process contains the following steps.[7]

1. The cross section of each slice layer is calculated based on the geometric model of the part and the desired layer thickness.

2. The optical mask is generated conforming to each cross section.

3. After leveling, the platform is covered with a thin layer of liquid photopolymer.

4. The mask corresponding to the current layer is positioned over the surface of the liquid resin, and the resin is exposed to a high-power UV lamp.

5. The residual liquid is removed from the workpiece by an aerodynamic wiper.

6. A layer of melted wax is spread over the workpiece to fill voids. The wax is then solidified by applying a cold plate to it.

7. The layer surface is trimmed to the desired thickness by a milling disk.

8. The current workpiece is covered with a thin layer of liquid polymer and step 4 to 7 are repeated for each succeeding upper layer until the topmost layer has been processed.

9. The wax is melted away upon completion of the part.

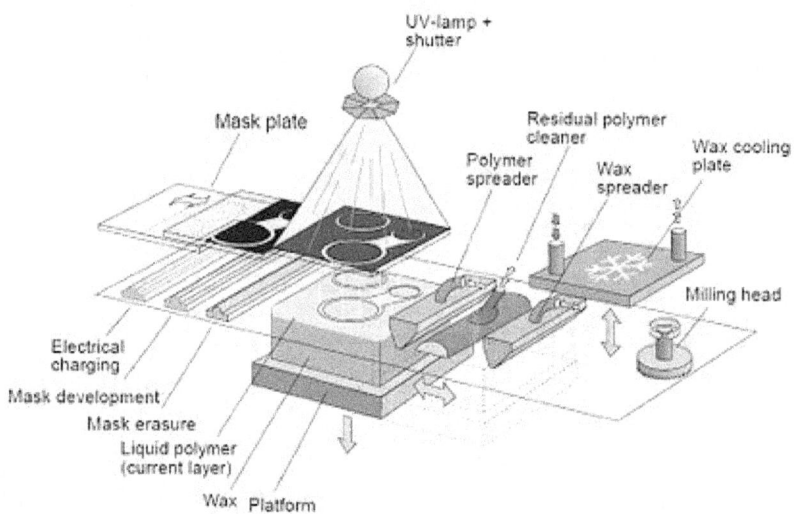

Schematic Diagram of Solid Ground Curing Process

63.2   Advantages and disadvantages

The primary advantage of the solid ground curing system is that it does not require a support structure since wax is used to fill the voids.[8] The model produced by SGC process is comparatively accurate in the Z-direction because the layer is milled after each light-exposure process.[9] Although it offers good accuracy coupled with high throughput, it produces too much waste and its operating costs are comparatively high due to system complexity.[10]

63.3   References

[1] The engineer: The rise of additive manufacturing(n.d.). Retrieved from

[2] Gebhardt, I.A.(2003). Rapid Prototyping: Industrial Rapid Prototyping System: Prototyper: Solid Ground Curing – Cubital. (pp. 105-109)

[3] Solid Ground Curing(n.d.). Retrieved from

[4] Castle Island Co.. (2002, June 22). Solid Ground Curing. Retrieved from

[5] Gebhardt, I.A.(2003). Rapid Prototyping: Industrial Rapid Prototyping System: Prototyper: Solid Ground Curing – Cubital. (pp. 105-109)

[6] Rapid Prototyping: Rapid Ground Curing(n.d.). Retrieved from

[7] Lee, K.W. (1999). Principles of CAD/CAM/CAE Systems: Rapid Prototyping and Manufacturing: Solid Ground Curing (pp. 383-384).

[8] Dolenc, A.(1994).*An Overview Of Rapid Prototyping Technologies In Manufacturing:Solid Ground Curing.* (p. 8)

[9] Gebhardt, I.A.(2003). Rapid Prototyping: Industrial Rapid Prototyping System: Prototyper: Solid Ground Curing – Cubital. (pp. 105-109)

[10] Rapid Ground Curing: An Introduction(n.d.). Retrieved from

Chapter 64

Sprout (computer)

Sprout by HP (styled as **sprout**) is a personal computer from Hewlett-Packard announced on October 29, 2014 and released for sale on November 9, 2014. The system was conceived by Brad Short.[1][2] Eric Monsef, a former Apple, Inc employee, is leading the sprout team.[3]

Sprout has dual interactive screens - a vertical touch screen display positioned in-line above a horizontal 'TouchMat' display, and also down-facing imaging sensors. The "Sprout Illuminator" serves as a projector, camera, depth sensor and scanner, and projects an image, for example a virtual keyboard, on the TouchMat.[4] For additional input, Sprout includes an Adonit stylus, as well as a Bluetooth keyboard and mouse set. This enables users to interact with physical and digital content while working.[5] Content can be digitally captured and manipulated in 2D or 3D directly on the TouchMat interface.[5] HP claims that this greatly simplifies and streamlines the creative process.

64.1 References

[1] Jack Clark. "HP Unveils Cheaper, 3-D Printing System to Spur Sales". *Bloomberg.com*.

[2] ABC News. "Hewlett Packard's New Sprout Aims to Bridge Physical and Digital Worlds". *ABC News*.

[3] Edward C. Baig, USA TODAY (29 October 2014). "First Look: HP pushes into 3-D printing, Blended Reality". *USA Today*.

[4] "HP's Sprout is like no computer you've ever seen". *www.cbsnews.com*. CBS News. 2014-10-31. Retrieved 2 September 2015.

[5] "Hands-on with the HP Sprout, an imaging powerhouse built into a touch-friendly PC". PCWorld.com. 2015-02-03. Retrieved 2015-06-12.

64.2 External links

- Official website

Chapter 65

Stereolithography

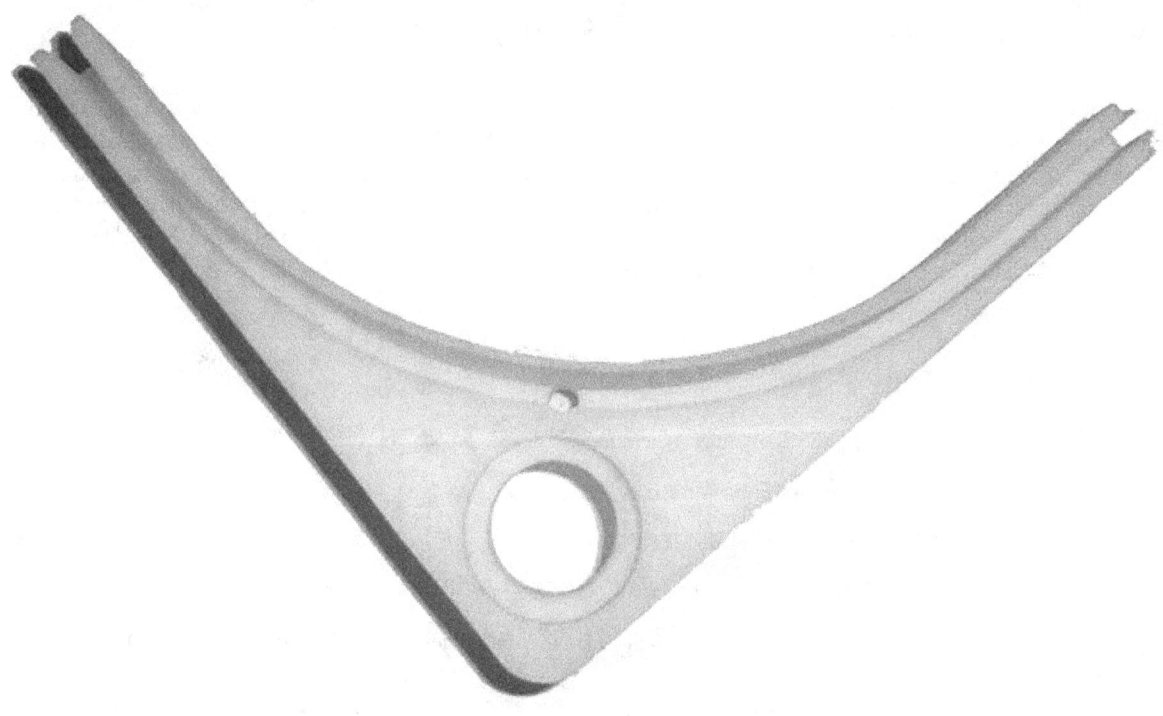

An SLA produced part

Stereolithography (**SLA** or **SL**; also known as **optical fabrication**, **photo-solidification**, **solid free-form fabrication**, **solid imaging** and **Resin printing**) is an additive manufacturing or 3D printing technology used for producing models, prototypes, patterns, and production parts up one layer at a time using lithographic methods.[1] For example by curing a photo-reactive resin with a UV laser or another similar power source.[2]

65.1 History

The term "stereolithography" was coined in 1986 by Charles (Chuck) W. Hull,[3] who patented it as a method and apparatus for making solid objects by successively "printing" thin layers of an ultraviolet curable material one on top of the other. Hull's patent described a concentrated beam of ultraviolet light focused onto the surface of a vat filled with liquid photopolymer. The light beam draws the object onto the surface of the liquid layer by layer, using polymerization or cross-linking to create a solid, a complex process which requires automation. In 1986, Hull founded the first company to generalize and commercialize this procedure, 3D Systems Inc,[4][5][6] which is currently based in Rock Hill, SC. More recently, attempts have been made to construct mathematical models of the stereolithography process and design algorithms to determine whether a proposed object may be constructed by the process.[7]

65.2   Technology

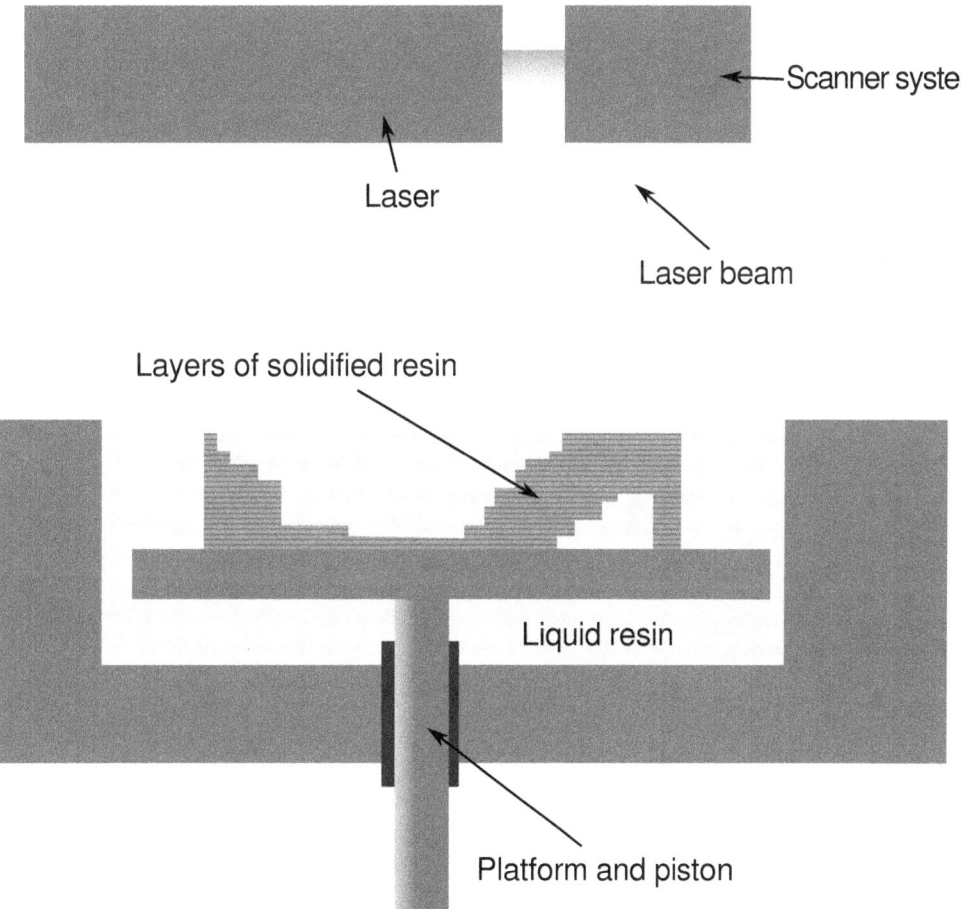

Stereolithography apparatus

Stereolithography is an additive manufacturing process which employs a vat of liquid ultraviolet curable photopolymer "resin" and an ultraviolet laser to build parts' layers one at a time. For each layer, the laser beam traces a cross-section of the part pattern on the surface of the liquid resin. Exposure to the ultraviolet laser light cures and solidifies the pattern traced on the resin and joins it to the layer below.

After the pattern has been traced, the SLA's elevator platform descends by a distance equal to the thickness of a single layer, typically 0.05 mm to 0.15 mm (0.002 in to 0.006 in). Then, a resin-filled blade sweeps across the cross section of the part, re-coating it with fresh material. On this new liquid surface, the subsequent layer pattern is traced, joining the previous layer. A complete 3-D part is formed by this process. After being built, parts are immersed in a chemical bath in order to be cleaned of excess resin and are subsequently cured in an ultraviolet oven.

Stereolithography requires the use of supporting structures which serve to attach the part to the elevator platform, prevent deflection due to gravity and hold the cross sections in place so that they resist lateral pressure from the re-coater blade. Supports are generated automatically during the preparation of 3D Computer Aided Design models for use on the stereolithography machine, although they may be manipulated manually. Supports must be removed from the finished product manually, unlike in other, less costly, rapid prototyping technologies.

65.3 Advantages and disadvantages

One of the advantages of stereolithography is its speed; functional parts can be manufactured within a day. The length of time it takes to produce one particular part depends on the size and complexity of the project and can last from a few hours to more than a day. Most stereolithography machines can produce parts with a maximum size of approximately 50×50×60 cm (20×20×24 in) and some, such as the Mammoth stereolithography machine (which has a build platform of 210×70×80 cm),[8] are capable of producing single parts of more than 2 m in length. Prototypes made by stereolithography are strong enough to be machined and can be used as master patterns for injection molding, thermoforming, blow molding, and various metal casting processes.

Although stereolithography can produce a wide variety of shapes, it has often been expensive; the cost of photo-curable resin has long ranged from $80 to $210 per liter, and the cost of stereolithography machines has ranged from $100,000 to more than $500,000.

Recently, renewed public interest in stereolithography has inspired the design of several consumer model printers with drastically reduced prices, such as the Titan 1 by Kudo3D, the Ilios HD by GizmoForYou, the Form 1 by Formlabs, the Pegasus Touch by FSL3D and the Nobel 1.0 by XYZPrinting. There has also been a drastic reduction in the cost of photo-curable resins, with USA based providers such as MakerJuice Labs offering materials as low as $55 per liter and European based providers such as spot-A Materials offering materials for €68 per liter.

65.4 See also

- Stereolithography (medicine)

- Thermoforming

65.5 References

[1] U.S. Patent 4,575,330 ("Apparatus for Production of Three-Dimensional Objects by Stereolithography")

[2] "How Stereolithography Works". THRE3D.com. Retrieved 4 February 2014.

[3] U.S. Patent 4,575,330 ("Apparatus for Production of Three-Dimensional Objects by Stereolithography")

[4] 3D Systems Inc Company Info

[5] Stereolithography

[6] What is Stereolithography?

[7] B. Asberg, G. Blanco, P. Bose, J. Garcia-Lopez, M. Overmars, G. Toussaint, G. Wilfong and B. Zhu, "Feasibility of design in stereolithography," *Algorithmica*, Special Issue on Computational Geometry in Manufacturing, Vol. 19, No. 1/2, Sept/Oct, 1997, pp. 61–83.

[8] Mammoth stereolithography: Technical specifications. materialise.com

65.6   Notes

- Kalpakjian, Serope and Steven R. Schmid. *Manufacturing Engineering and Technology* 5th edition. Ch. 20 (pp. 586–587 Pearson Prentice Hall. Upper Saddle River NJ, 2006.

65.7   External links

- Full Spectrum Laser rolls out two new innovative 3D Printers at 3D Printer World

- Ilios modular 3d Printer by GizmoForYou with industrial grade motion, fully metallic construction and highly accurate repeatability rates

- Super Precise Kudo3D Titan 1 SLA 3D Printer Hits Kickstarter on May 27 Starting at $1899

- Pegasus Touch by FSL3D Laser SLA 3D Printer: Low cost, High Quality on Kickstarter

- Formlabs Form 1 a low cost Stereolithography printer being designed and fabricated as a Kickstarter project

- How Stereolithography (3-D Layering) Works from HowStuffWorks.com

- Rapid Prototyping and Stereolithography animation – Animation demonstrates stereolithography and the actions of an SL machine

- Video of micro-stereolithography for biomedical applications

Chapter 66

Stereolithography (medicine)

Stereolithographic models have been used in medicine since the 1990s,[1] for creating 3D corporeal models of various anatomical regions of a patient, based on datasets from CT-scans.

66.1   Usage

- Stereolithography is a Rapid prototyping process that creates solid physical models directly from computer data. In industry this data comes from 3D computer-aided design (CAD) data. The process can also be used to build highly accurate replicas of human (or animal) anatomy by using computer images from medical scanners.[2] Typically Computed tomography (CT) is used but Magnetic Resonance Imaging (MRI) can also be used. Models have also been made from Ultra Sound and more recently from lower cost Cone Beam CT scanners. Medical models can also be made using a range of other Rapid Prototyping processes although stereolithography remains popular.

- Medical models are used in medicine and surgery to provide surgeons with a better appreciation of the anatomical situation of a patient, before surgery. Although the advent of improved 3D computer reconstruction and virtual surgical planning means that in some cases models are not needed they remain popular for complex surgeries particularly in cranial surgery, maxillofacial surgery, oral surgery and neurosurgery.

- Stereolithographic models are used as an aid to diagnosis, preoperative planning and implant design and manufacture. This might involve for example planning and rehearsing osteotomies. Surgeons use models to help plan surgeries but prosthetists and technologists also use models as an aid to the design and manufacture of custom-fitting implants. Medical models are frequently used to help in the construction of Cranioplasty plates for example.

66.2   Medical Modelling Process

The process of medical modelling involves several stages including image acquisition, image segmentation, data translation, model building and post-processing.[3] Medical modelling involves first acquiring a 3D CT scan (or other form of scan data). The CT data should be in a suitable format and acquired using suitable parameters to obtain a high quality model.[4] This data consists of a series of cross sectional images of the human anatomy. In these images different tissues show up as different levels of grey. Selecting a range of grey values enables specific tissues to be isolated. A region of interest is then selected and all the pixels connected to the target point within that grey value range are selected. This enables a specific organ to be selected. Most frequently this will be bone but it could be any tissue that can be identified in the scan image. This process is referred to as segmentation. The segmented data may then be interpolated and have other processes performed on it to translate it into a format suitable for the stereolithography process.

Whilst the stereolithography process is inherently accurate the accuracy of a medical model depends on many factors, especially the operator performing the segmentation correctly. There are potential errors possible when making medical models using stereolithography but these are easy to avoid with practice and well trained operators.[5]

66.3   Commercial Services

There are several specialist companies that provide medical modelling services such as for example PDR in the United Kingdom, Medical Modeling Inc. in the USA and Materialise in Belgium. As Rapid Prototyping machines become more affordable many hospitals are investing in their own medical modelling facilities.

66.4   References

[1] Klimek, L; Klein HM; Schneider W; Mosges R; Schmelzer B; Voy ED (1993). "Stereolithographic modelling for reconstructive head surgery". *Acta Oto-Rhino-Laryngologica Belgica* **47** (3): 329–34.

[2] Bouyssie, JF; Bouyssie S; Sharrock P; Duran D (1997). "Stereolithographic models derived from x-ray computed tomography. Reproduction accuracy". *Surgical & Radiologic Anatomy* **19** (3): 193–9.

[3] Bibb, Richard (2006). *Medical Modelling: the application of advanced design and development technologies in medicine.* Cambridge: Woodhead Publishing Ltd. ISBN 1-84569-138-5.

[4] Winder, RJ; Bibb, R (2009). "A Review of the Issues Surrounding Three-Dimensional Computed Tomography for Medical Modelling using Rapid Prototyping Techniques". *Radiography* **16**: 78–83. doi:10.1016/j.radi.2009.10.005.

[5] Winder, RJ; Bibb, R (2005). "Medical Rapid Prototyping Technologies: State of the Art and Current Limitations for Application in Oral and Maxillofacial Surgery". *Journal of Oral and Maxillofacial Surgery* **63** (7): 1006–15. doi:10.1016/j.joms.2005.03.016.

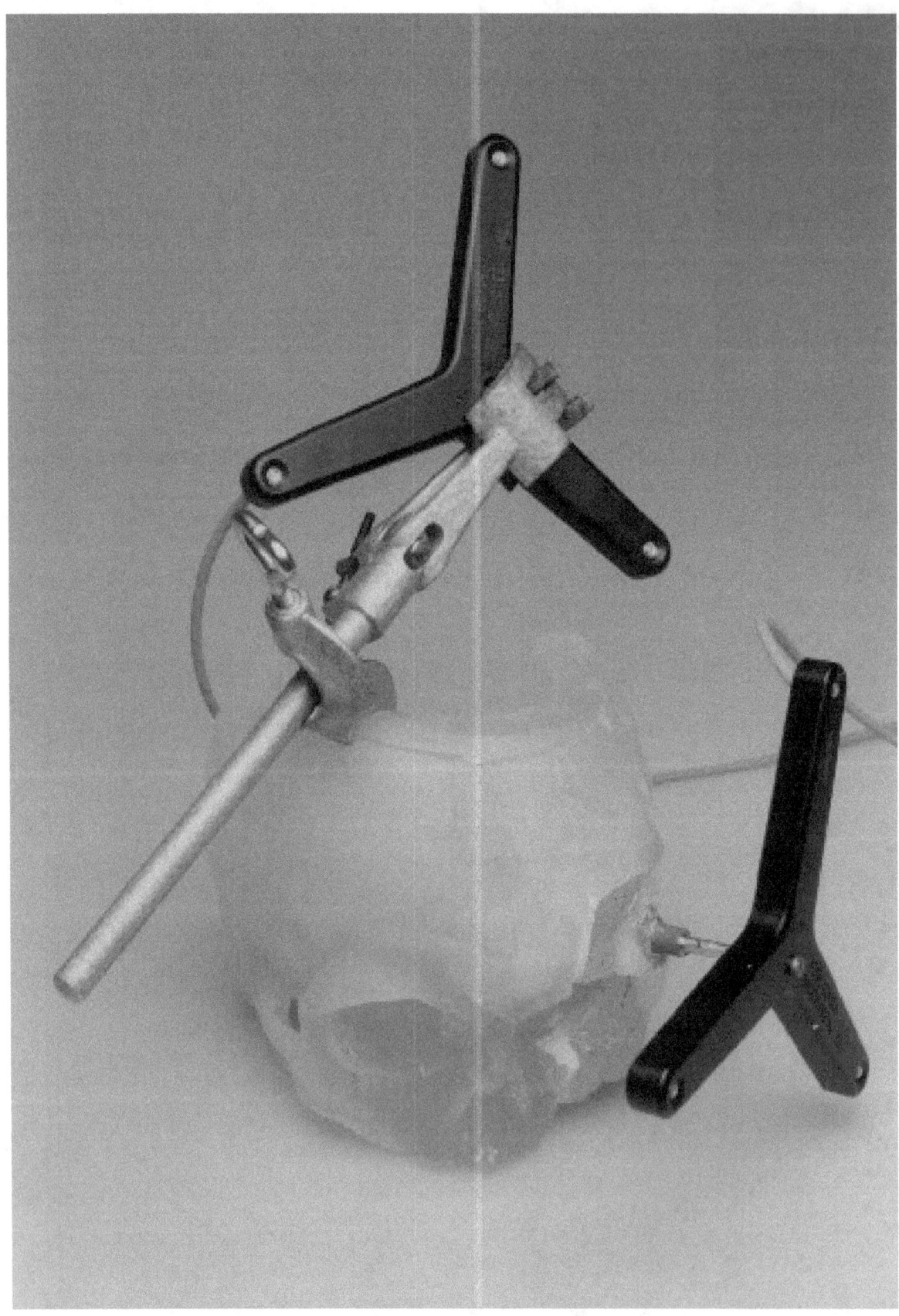

Stereolithographic model of a skull, using an infrared system

Chapter 67

STL (file format)

STL (**STereoLithography**) is a file format native to the stereolithography CAD software created by 3D Systems.[1][2][3] STL has several after-the-fact backronyms such as "Standard Triangle Language" and "Standard Tessellation Language".[4] This file format is supported by many other software packages; it is widely used for rapid prototyping, 3D printing and computer-aided manufacturing.[5] STL files describe only the surface geometry of a three-dimensional object without any representation of color, texture or other common CAD model attributes. The STL format specifies both ASCII and binary representations. Binary files are more common, since they are more compact.[6]

An STL file describes a raw unstructured triangulated surface by the unit normal and vertices (ordered by the right-hand rule) of the triangles using a three-dimensional Cartesian coordinate system. STL coordinates must be positive numbers, there is no scale information, and the units are arbitrary.[7]

67.1 ASCII STL

An ASCII STL file begins with the line

solid *name*

where *name* is an optional string (though if *name* is omitted there must still be a space after solid). The file continues with any number of triangles, each represented as follows:

facet normal n_i n_j n_k outer loop vertex $v1x$ $v1y$ $v1z$ vertex $v2x$ $v2y$ $v2z$ vertex $v3x$ $v3y$ $v3z$ endloop endfacet

where each n or v is a floating-point number in sign-mantissa-"e"-sign-exponent format, e.g., "2.648000e-002" (noting that each v must be non-negative). The file concludes with

endsolid *name*

The structure of the format suggests that other possibilities exist (e.g., facets with more than one "loop", or loops with more than three vertices). In practice, however, all facets are simple triangles.

White space (spaces, tabs, newlines) may be used anywhere in the file except within numbers or words. The spaces between "facet" and "normal" and between "outer" and "loop" are required.[6]

67.2 Binary STL

Because ASCII STL files can become very large, a binary version of STL exists. A binary STL file has an 80-character header (which is generally ignored, but should never begin with "solid" because that will lead most software to assume that this is an ASCII STL file). Following the header is a 4-byte unsigned integer indicating the number of triangular facets in the file. Following that is data describing each triangle in turn. The file simply ends after the last triangle.

Each triangle is described by twelve 32-bit floating-point numbers: three for the normal and then three for the X/Y/Z coordinate of each vertex – just as with the ASCII version of STL. After these follows a 2-byte ("short") unsigned integer that is the "attribute byte count" – in the standard format, this should be zero because most software does not understand anything else.[6]

Floating-point numbers are represented as IEEE floating-point numbers and are assumed to be little-endian, although this is not stated in documentation.

UINT8[80] – Header UINT32 – Number of triangles
foreach triangle REAL32[3] – Normal vector REAL32[3] – Vertex 1 REAL32[3] – Vertex 2 REAL32[3] – Vertex 3
UINT16 – Attribute byte count end

67.3   Color in binary STL

There are at least two non-standard variations on the binary STL format for adding color information:

- The VisCAM and SolidView software packages use the two "attribute byte count" bytes at the end of every triangle to store a 15-bit RGB color:

  - bit 0 to 4 are the intensity level for blue (0 to 31),
  - bits 5 to 9 are the intensity level for green (0 to 31),
  - bits 10 to 14 are the intensity level for red (0 to 31),
  - bit 15 is 1 if the color is valid, or 0 if the color is not valid (as with normal STL files).

- The Materialise Magics software uses the 80-byte header at the top of the file to represent the overall color of the entire part. If color is used, then somewhere in the header should be the ASCII string "COLOR=" followed by four bytes representing red, green, blue and alpha channel (transparency) in the range 0–255. This is the color of the entire object, unless overridden at each facet. Magics also recognizes a material description; a more detailed surface characteristic. Just after "COLOR=RGBA" specification should be another ASCII string ",MATERIAL=" followed by three colors (3×4 bytes): first is a color of diffuse reflection, second is a color of specular highlight, and third is an ambient light. Material settings are preferred over color. The per-facet color is represented in the two "attribute byte count" bytes as follows:

  - bit 0 to 4 are the intensity level for red (0 to 31),
  - bits 5 to 9 are the intensity level for green (0 to 31),
  - bits 10 to 14 are the intensity level for blue (0 to 31),
  - bit 15 is 0 if this facet has its own unique color, or 1 if the per-object color is to be used.

The red/green/blue ordering within those two bytes is reversed in these two approaches – so while these formats could easily have been compatible, the reversal of the order of the colors means that they are not – and worse still, a generic STL file reader cannot automatically distinguish between them. There is also no way to have facets be selectively transparent because there is no per-facet alpha value – although in the context of current rapid prototyping machinery, this is not important.

67.4   The facet normal

In both ASCII and binary versions of STL, the **facet normal** should be a unit vector pointing outwards from the solid object. In most software this may be set to (0,0,0), and the software will automatically calculate a normal based on the order of the triangle vertices using the "right-hand rule". Some STL loaders (e.g. the STL plugin for Art of Illusion) check that the normal in the file agrees with the normal they calculate using the right-hand rule and warn the user when it does not. Other software may ignore the facet normal entirely and use only the right-hand rule. Although it is rare to specify a normal that cannot be calculated using the right-hand rule, in order to be entirely portable, a file should both provide the facet normal and order the vertices appropriately. A notable exception is SolidWorks, which uses the normal for shading effects.

67.5   History of use

Stereolithography machines are 3D printers that can build any volume shape as a series of slices. Ultimately these machines require a series of closed 2D contours that are filled in with solidified material as the layers are fused together. A natural file format for such a machine would be a series of closed polygons corresponding to different Z-values. However, since it is possible to vary the layer thicknesses for a faster though less precise build, it was easier to define the model to be built as a closed polyhedron that can be sliced at the necessary horizontal levels.

The STL file format appears capable of defining a polyhedron with any polygonal facet, but in practice it is only ever used for triangles, which means that much of the syntax of the ASCII protocol is superfluous.

To properly form a 3D volume, the surface represented by any STL files must be closed and connected, where every edge is part of exactly two triangles, and not self-intersecting. Since the STL syntax does not enforce this property, it can be ignored for applications where the closedness does not matter. The closedness only matters insofar as the software that slices the triangles requires it to ensure that the resulting 2D polygons are closed. Sometimes such software can be written to clean up small discrepancies by moving vertices that are close together so that they coincide. The results are not predictable, but it is often sufficient.

67.6   Use in other fields

STL file format is simple and easy to output. Consequently, many computer-aided design systems can output the STL file format. Although the output is simple to produce, some connectivity information is discarded.

Many computer-aided manufacturing systems require triangulated models. STL format is not the most memory- and computationally efficient method for transferring this data, but STL is often used to import the triangulated geometry into the CAM system. The format is commonly available, so the CAM system will use it. In order to use the data, the CAM system may have to reconstruct the connectivity.

STL can also be used for interchanging data between CAD/CAM systems and computational environments such as Mathematica.

67.7   Notes

On Microsoft Windows, the .stl file extension is used for Certificate Trust Lists; file listings will therefore mark stereolithography files as Certificate Trust Lists. Windows 10 recognizes the .stl file extension as "3D-Object".

67.8   See also

- Additive Manufacturing File Format (AMF), a newer standard with native support for color, multiple materials, and constellations

- Clara.io, a free online 3D editor that can import, edit, and export STL files

- CloudCompare, an open-source application for handling STL files

- Mathematica, a technical computing system that can work with STL files

- MeshLab, a free and open-source cross-platform application for visualizing, processing, and converting three-dimensional meshes to or from the STL file format

- PLY (file format), an alternative file format offering more flexibility than most stereolithography applications

- Wavefront .obj file, a 3D geometry definition file format with *.obj* file extension

- X3D, a royalty-free ISO standard for 3D computer graphics

67.9 References

[1] *StereoLithography Interface Specification*, 3D Systems, Inc., July 1988

[2] *StereoLithography Interface Specification*, 3D Systems, Inc., October 1989

[3] *SLC File Specification*, 3D Systems, Inc., 1994

[4] Grimm, Todd (2004), *User's Guide to Rapid Prototyping*, Society of Manufacturing Engineers, p. 55, ISBN 0-87263-697-6. Many names are used for the format: for example, "standard triangle language", "stereolithography language", and "stereolithography tesselation language". Page 55 states, "Chuck Hull, the inventor of stereolithography and 3D Systems' founder, reports that the file extension is for stereolithography."

[5] Chua, C. K; Leong, K. F.; Lim, C. S. (2003), *Rapid Prototyping: Principles and Applications* (2nd ed.), World Scientific Publishing Co, ISBN 981-238-117-1 Chapter 6, Rapid Prototyping Formats. Page 237, "The STL (STeroLithography) file, as the de facto standard, has been used in many, if not all, rapid prototyping systems." Section 6.2 STL File Problems. Section 6.4 STL File Repair.

[6] Burns, Marshall (1993). *Automated Fabrication*. Prentice Hall. ISBN 978-0-13-119462-5.

[7] Fabbers.com, The StL Format: Standard Data Format for Fabbers, reprinted from Marshall Burns, Automated Fabrication, http://www.ennex.com/~{}fabbers/StL.asp stating, "The object represented must be located in the all-positive octant. In other words, all vertex coordinates must be positive-definite (nonnegative and nonzero) numbers. The StL file does not contain any scale information; the coordinates are in arbitrary units."

67.10 External links

- The StL Format: Standard Data Format for Fabbers

- File Extension STL: List of software to work with STL file

Chapter 68

Strati (automobile)

Overall view of Strati[1]

Strati is an electric car developed by Local Motors and manufactured in collaboration with Cincinnati Incorporated and Oak Ridge National Laboratory.[2] It is the world's first 3D-printed electric car.[3] The car was manufactured using a Large Scale 3D Printer developed by ORNL and Cincinnati Inc. The car took just 44 hours to print during the 2014 International Manufacturing Technology Show in Chicago, Illinois. The printing was followed by three days of milling and assembling, with the completed car first test-driven on September 13, 2014. Strati is claimed to be the world's first 3D-Printed electric car.[1][4]

68.1 Design

In April 2014, Local Motors organized the 3D Printed Car Design Challenge crowdsourcing to assist in the production of a full-body 3D-printed car. Seven finalists were selected from more than 200 submissions. In June 2014, Local Motors announced that the challenge was won by Michele Anoé of Italy, who was awarded the $5,000 prize.[5] After the contest, Local Motors took the design and made several modifications so that the car could be manufactured through 3D-Printing.

68.2 Specifications

The two-seat Strati is considered to be a "neighborhood" electric car. Depending on the configuration of the battery packs, the range of the car can be 100 to 120 mi (160 to 190 km) with top speeds of 40 mph (64 km/h). The car is not designed to be used on highways, as it does not meet the required safety test requirements. Production is planned by the end of 2015, with prices between $18,000 and $30,000.[6]

- Front view with steering details exposed

- Passenger side

- Rear view

68.3 Manufacturing

Following the design competition, Local Motors handed off the design to the engineers at ORNL who perfected the process of Large Scale 3D Printing such that the Local Motors design could actually be manufactured. ORNL worked with Cincinnati Incorporated to develop the printer that would allow for the printing of the entire car. With the printer, ORNL and Cincinnati Inc. manufactured all body parts of the car and allowed for easy mounting of the mechanical parts, such as the electric motors and batteries.

Strati is printed from thermoplastic using a big area additive manufacturing (BAAM)[8] machine (this is, a big area 3D-printing machine). This material is fully recyclable, which can be chopped and reprocessed to be used in printing another car. After the car is printed, the mechanical and electrical parts such as battery, motors, and suspension are manually assembled.[6]

The printing process has been improved by ORNL since July 2014, bringing the printing time of 140 hours down to less than 45 hours in September. Since IMTS, ORNL has brought the printing time of the Strati to less than 24 hours and is continuing their research efforts with the hope of printing the car in less than 10 hours.

68.4 The world's first title

Disputes exist over the title of the world's first 3D-printed car. In 2010, a hybrid car "Urbee"[9] was 3D-printed using an additive manufacturing process for the entire body.[10] Local Motors claimed that Urbee's manufacturer only 3D-printed the panels and other exterior parts, but used standard parts for the internal structure. For Strati, the company claimed that 3D printing was used for all except the parts that are "mechanically involved". Strati still holds the honor of the world's first 3D-printed electric car.[1][3]

68.5 See also

- Local Motors

Details of the printed body of a Strati[7]

68.6   References

[1] Robarts, Stu (17 September 2014). ""World's first" 3D printed car created and driven by Local Motors". *Gizmag*. Retrieved 22 September 2014.

[2] Gastelu, Gary (3 July 2014). "Local Motors 3D-printed car could lead an American manufacturing revolution". Fox News. Retrieved 22 September 2014.

[3] Russon, Mary-Ann (16 September 2014). "The Strati: World's First 3D-Printed Electric Car Built in Just 44 Hours". *IB Times*. Retrieved 22 September 2014.

[4] Franklin, Dallas (15 September 2014). "Made in Chicago: World's First 3D Printed Electric Car". KFOR-TV. Retrieved 22 September 2014.

[5] Jeffrey, Colin (9 June 2014). "Strati wins 3D printed car challenge". *Gizmag*. Retrieved 22 September 2014.

[6] Pyper, Julia (12 September 2014). "World's First Three-Dimensional Printed Car Made in Chicago". *Scientific American*. Retrieved 22 September 2014.

[7] France, Anna Kaziunas (20 September 2014). "First Fused-Filament, Fully-Electric Vehicle". *Makezine*. Retrieved 22 September 2014.

[8] BAAM.

[9] Urbee.

[10] Quick, Darren (2 November 2010). "The Urbee hybrid: the world's first 3D printed car". *Gizmag*. Retrieved 22 September 2014.

68.7 External links

- Media related to Strati automobiles at Wikimedia Commons
- Local Motors Website
- Cincinnati Incorporated

Chapter 69

Thingiverse

Thingiverse is a website dedicated to the sharing of user-created digital design files. Providing primarily open source hardware designs licensed under the GNU General Public License or Creative Commons licenses, users choose the type of user license they wish to attach to the designs they share. 3D printers, laser cutters, milling machines and many other technologies can be used to physically create the files shared by the users on Thingiverse.

Thingiverse is widely used in the DIY technology and Maker communities, by the RepRap Project, and by 3D Printer and MakerBot operators. Numerous technical projects use Thingiverse as a repository for shared innovation and dissemination of source materials to the public. Many of the objects are for the purpose of repair.[1]

69.1 History

Thingiverse was started in November 2008[2] by Zach Smith as a companion site to MakerBot Industries, a DIY 3D printer kit making company.

Thingiverse received an Honorable Mention in the Digital Communities category of the 2010 ARS Electronica | Prix Ars Electronica international competition for cyber-arts.[3]

There were 25,000 designs uploaded to Thingiverse as of November 2012[4] and more than 100,000 in June 2013.[5] The 400000th Thing was published on the 19 July 2014[6]

69.2 Administration

The site is owned by MakerBot Industries and run by one of its founders, Bre Pettis in Brooklyn, New York.

Thingiverse terms of use include the agreement that users don't include content that "contributes to the creation of weapons, illegal materials or is otherwise objectionable".[7]

69.3 Open source hardware

Where many open source hardware projects are focused on project-specific materials, Thingiverse provides a common ground from which derivatives[8] and mashups[9] can form.

Thingiverse contains many improvements and modifications that are generated by the community that surrounds open source hardware. There are numerous files to open-source improve, upgrade and modify RepRap and Contraptor 3D printers.

69.4 See also

- .dwg

- Pinshape

- Sketchfab

- Printolia

- 3D Printing Marketplace

- Materialise NV

- 3DLT

- Sculpteo

- Shapeways

- Threeding

- STL (file format)

69.5 References

[1] "Make and Mend: Thingiverse fixit roundup, Makezine.com by John Baichtal, 16 August 2010". Blog.makezine.com. 2010-08-16. Retrieved 2011-09-16.

[2] Previous post Next post (2008-11-20). "Thingiverse.com Launches A Library of Printable Objects, Wired; GeekDad by John Baichtal, November 20, 2008". Wired.com. Retrieved 2011-09-16.

[3] Austria. "2010 ARS Electronica | Prix Ars Electronica | Digital Communities | ANERKENNUNGEN". New.aec.at. Retrieved 2011-09-16.

[4] "Introducing MakerBot Thingiverse Dashboard And Follow Features". Makerbot blog. |first1= missing |last1= in Authors list (help)

[5] "The 100,000th Thing on Thingiverse!". Makerbot blog. |first1= missing |last1= in Authors list (help)

[6] 400 000th thing on Thingiverse

[7] "Daily Dot".

[8] "Prusa simplified mendel by prusajr". Thingiverse.com. 2010-09-18. Retrieved 2011-09-16.

[9] "Duplo Brick to Brio Track adapter with snap-lock by Zydac". Thingiverse.com. Retrieved 2011-09-16.

69.6 External links

- Official website

Chapter 70

Threeding

Threeding is an online 3D Printing Marketplace and community for trading and free exchange of files ready for 3D printing. The website follows in the tradition of open marketplace, giving sellers personal storefronts where they list their 3D printable models and make them available for the global audience. **Threeding** is one of the several 3D files repositories that have emerged with the fast-growing 3D printing industry.[1]

70.1 History

Threeding was started in 2013 by a group of students from the Bulgarian National Academy of Arts. The website quickly became very popular among CAD designers, hobbyists and tech geeks. Significant parts of the 3D objects available at **Threeding** are digital copies of historical artifacts which is a new line in the world of 3D printing repositories. Several Eastern European historical and archaeological museums have also opened stores and sell 3D printable models of their exhibits via Threeding.[2][3][4]

70.2 Concept

Threeding works in a manner similar to early eBay. In order to sell 3D printing models on Threeding, users must register and thus, create a store. Creating a store and uploading a product on **Threeding** is free, however the website charges a commission for each sale. The website does not charge anything for free sharing of 3D models. Buying 3D printing models is intuitive and similar to any other eCommerce website: a buyer browses through the website and once finds a model, he/she click Add to Cart and proceed to Check-out. The website has integrated Paypal and the major Credit cards.[5]

70.3 3D Printing Cultural Heritage

Several Eastern European historical and archaeological museums have signed cooperation agreements with **Threeding** and opened virtual stores. **Threeding** 3D scans museums' exhibits and the museums sell the digital 3D scans through their virtual stores. The available 3D printing models are historical artifacts from the prehistoric period, ancient time, medieval and the modern history.[6][7][8][9]

70.4 References

[1] "Sharing and paid exchange of 3D printable files", 3D Printer Cafe, Jan 10, 2014

[2] "Printable Heritage: How Threeding is 3D Printing History", THRE3D, March 20, 2014

[3] "Bulgaria Feels the Benefit of History With Three-d-(print)ing Cooperation", 3D Printing Industry, March 6, 2014

[4] "3D printing marketplace Threeding re-launches with fresh new look"

[5] "In Bulgaria, an eBay for 3D Printable Designs Called Threeding Emerges", On 3D Printing, January 11, 2014

[6] "Museum Sells Its Collection - As 3D Prints", Fabbaloo, April 7, 2014

[7] "Bulgarian 3D model Marketplace partners with Regional Museum", 3DERS, April 3, 2014

[8] "Bulgarian 3D Printing Marketplace Teams Up with Regional Museum", On 3D Printing, March 4, 2014

[9] "Rare Museum Artifacts, Now Available for Purchase and Print at 3D Design Marketplace, Threeding"

70.5 External links

- Official website

Chapter 71

TRI-D (rocket engine)

TRI-D is a 3D printed metal rocket engine.[1][2] University of California, San Diego (UCSD) students built the metal rocket engine using a technique previously confined to NASA, using a GPI Prototype and Manufacturing Services printer[1] via the Direct metal laser sintering (DMLS) method.[1] UCSD students were the first group in the world to 3D print a rocket engine of its size, other than NASA as of February 2014.[3] The Tri-D engine cost US$6,800.[1][4][5]

71.1 Development

The Tri-D rocket engine was designed and built with the cooperation of NASA's Marshall Space Flight Center, to explore the feasibility of printed rocket components. It was designed to power the third stage of a Nanosat or Cubesat launcher, i.e. an engine capable of launching satellites that weigh less than 1.33 kg (2.93 lb).[1][2][5]

71.2 Specifications

Tri-D is around 17.7 cm long and weighs around 4.5 kg. It was fabricated using a chromium-cobalt alloy powder. The propellants are kerosene and liquid oxygen. The engine produces about 200 pounds-force (890 newtons; 91 kilograms-force) thrust. According to Gizmag "the injector has a Fuel-Oxidizer-Oxidizer-Fuel inlet arrangement with two outer fuel orifices converging with two inner oxidizer orifices".[1][4]

The engine has a regenerative cooling jacket that extends to the nozzle to prevent the engine from overheating while firing. The combustion chamber was designed to burn the propellants in the middle of the chamber and keep as much heat generated as possible away from its chamber walls, while at the same time insulating the wall with a film of cooler gases.[1]

71.3 Printer

The engine was printed with a GPI Prototype and Manufacturing Services printer using a technique called Direct metal laser sintering (DMLS). In the process of printing, a powder of the chromium-cobalt alloy is spread in a thin layer. Then computer-controlled laser fuses the powders into a cross section of the engine component. The machine then spreads a second layer of powder and the process continuously repeats until teach component is complete. Any excess powder is removed as are temporary supports that were printed to hold the components together during printing process. Finally it is hardened, polished and assembled.[1][4]

71.4 Test firing

The test firing at Mojave went without any problems and the engine exhaust achieved 200 pounds-force (890 newtons; 91 kilograms-force) thrust. The team claimed "it was a resounding success and could be the next step in the development of cheaper propulsion systems and a commercializing of space".[1]

71.4.1 Injector test

> On a separate engine, a 3D printed injector was test fired in a conventionally manufactured engine. In the test of the injector on August 22, 2014, the engine generated 20,000 pounds-force (89,000 newtons; 9,100 kilograms-force) thrust.[2]

71.5 Vulcan-I

The group is working on another rocket engine named Vulcan-I, nick-named "Tri-D's big brother".[6]

71.6 References

[1] UCSD students test fire 3D-printed metal rocket engine, gizmag, October 12, 2013. (archive)

[2] 3D-Printed Rocket Engine Built By Students Passes Big Test (Video), Space.com, October 08 2013. (archive)

[3] UCSD group plans second 3-D printed rocket engine test to gather more data - 10News.com KGTV ABC10 San Diego, 10News, February 25, 2014. (archive)

[4] University students successfully tested 3D printed rocket engine, 3Ders, October 7, 2013. (archive)

[5] Students successfully hot test 3D printed rocket engine, design-engineering, October 10, 2013. (archive)

[6] PROJECTS, USCD PROJECTS- SEDS. (archive)

Chapter 72

Ultrasonic consolidation

Ultrasonic Consolidation (**UC**), sometimes referred to as **Ultrasonic Additive Manufacturing** (**UAM**), is an additive manufacturing technique based on the ultrasonic welding of metal foils and CNC contour milling.[1] High-frequency (typically 20,000 hertz) ultrasonic vibrations are locally applied to metal foil materials, held together under pressure, to create a solid-state weld. CNC contour milling is then used to create the required shape for the given layer. This process is then repeated until a solid component has been created or a feature repaired/added to a component. UC has the ability to join dissimilar metal materials[2] of different thicknesses and allows the embedment of fibre materials at relatively low temperature, (typically less than 50% of the metal matrix melting temperature), and pressure into solid metal matrices.[3][4]

72.1   History

The Ultrasonic Consolidation process was invented and patented[5] by Dawn White. In 1999, White founded Solidica Inc.[6] which is the commercial owner and provider of the UC technology.[7] The commercial equipment for UC is called the Form-ation machine.

72.2   Process

As with most other additive manufacturing processes UC creates objects directly from a CAD model of the required object. The file is then "sliced" into layers which results in the production of a .STL file that can be used by the UC machine to build the required object, layer by layer.

The general manufacturing process is:[8]

- A base plate is placed onto the machine anvil and fixed into place.

- Metal foil is then drawn under the sonotrode, which applies pressure through a normal force and the ultrasonic oscillations, and bonded to the plate.

- This process is then repeated until the required area has been covered in ultrasonically consolidated material.

- A CNC mill is then used to trim the excess foil from the component and achieve the required geometry.

- The deposit and trim cycle is repeated until a specified height is reached, (typically 3–6 mm).

- At this height a smaller finishing mill is used to create the required tolerance and surface finish of the part.

- The deposit, trim and finish cycle continues until the finished object has been manufactured; at which point it is taken off the anvil and the finished article is removed from the base plate.

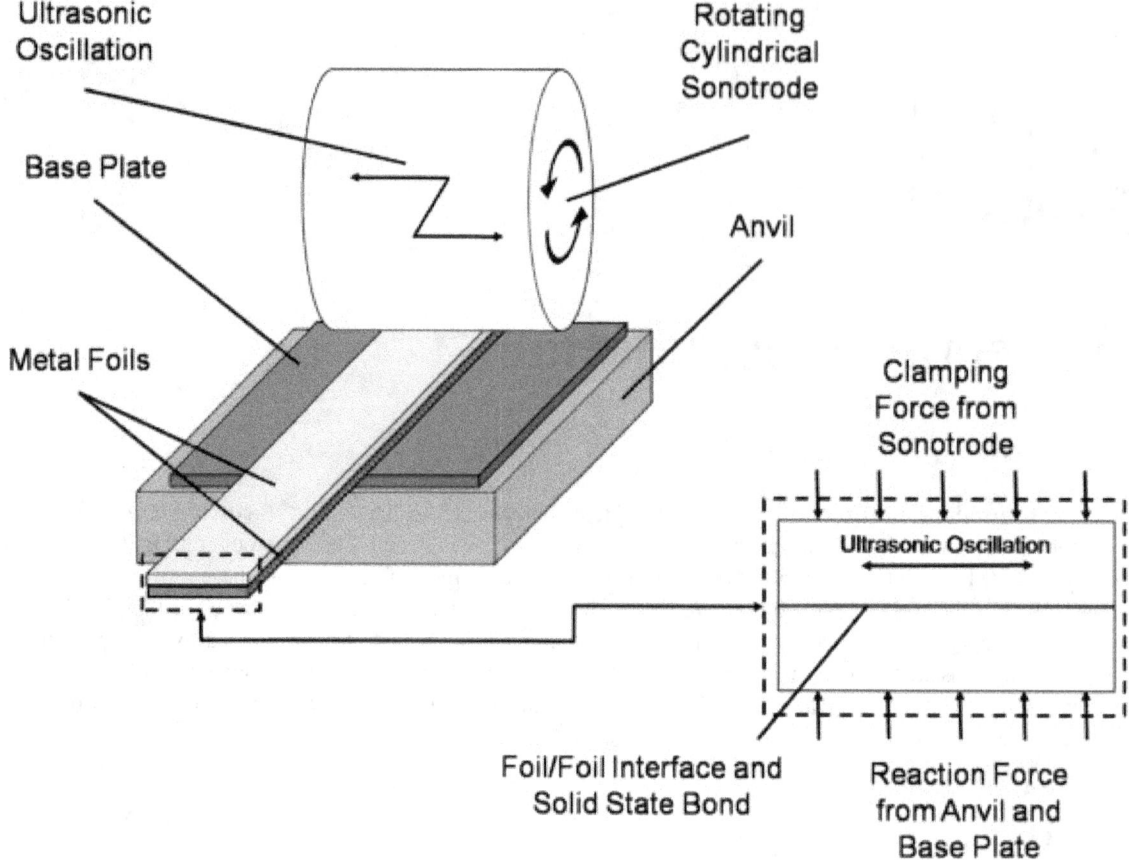

A schematic of the Ultrasonic Consolidation process.

72.3 References

[1] Advanced Materials and Processes, *Ultrasonic Consolidation of Aluminum Tooling*, D.R. White, Vol. 161, 2003, pp. 64–65

[2] Rapid Prototyping Journal, *Use of Ultrasonic Consolidation for Fabrication of Multi-Material Structures*, G.D. Janaki Ram; C. Robinson; Y. Yang; B.E. Stucker, Vol. 13, No. 4, 2007, pp. 226–235

[3] Composite Structures, *Ultrasonic Consolidation for Embedding SMA Fibres within Aluminium Matrices*, C.Y. Kong; R.C. Soar; P.M. Dickens, Vol. 66, No. 1–4, 2004, pp. 421–427

[4] Journal of Engineering Materials and Technology, *Characterization of Process for Embedding SiC Fibers in Al 6061 O Matrix Through Ultrasonic Consolidation*, D. Li; R.C. Soar, Vol. 131, No. 2, 2009, pp. 021016-1 to 021016-6

[5] http://www.freepatentsonline.com/6519500.html

[6] http://www.solidica.com

[7] http://home.att.net/~{}castleisland/tl_221b.htm

[8] http://www.solidica.com/systems.advanced.html

Chapter 73

Voxeljet

voxeljet AG, which is based in Friedberg (Bayern) near Augsburg (Germany), is a manufacturer of industrial 3D printing systems. The company has been listed on the New York Stock Exchange since its IPO in the year 2013.[3][4] Besides the development and distribution of printing systems, voxeljet AG also operates service centers for the on-demand manufacture of molds and models for metal casting in Germany and abroad. These products are manufactured with the help of a generative production method based on 3D CAD data (also referred to as "3D printing").

73.1 History

73.1.1 Beginnings

voxeljet AG traces its roots back to the year 1995, with the first successful drop-dosing of UV adhesives. The first 3D printing trials were conducted at the Precision Engineering department of the Technical University Munich as part of the "Generation of 3D structures" project. In 1996, Dr. Ingo Ederer participated in the first Munich business plan competition and was awarded his first patent in 1998. The first sand molds were printed in the same year.[5]

73.1.2 Formation of the company

Generis GmbH, the predecessor of today's voxeljet AG, was founded on 5 May 1999 by Ingo Ederer, Rainer Höchsmann and engineer Joachim Heinzl at Munich's Technical University.[6] The purpose of the company was the development of new generative processes for the production of cast and plastic components.[7] The company started its operations at Technical University Munich with four employees. Shortly afterwards, it refurbished and relocated into the premises in Augsburg.[5] In the year 2002, the company completed its first orders for the delivery of sand-based printers to BMW AG and Daimler AG, before opening the service center in Augsburg in the year 2003. In the same year, Bayern Kapital GmbH,[8] the Startkapital Fonds Augsburg and Franz Industriebeteiligungen AG[9] joined as new shareholders.

73.1.3 Establishment phase

The first VX800 system was sold to Alphaform AG in 2005.[5] This was followed by the sale of the first VX500 to the University of Rostock two years later.[10] In 2008, voxeljet technology GmbH received an award as part of the Bavarian Innovation Prize from the hands of Bavarian premier Günther Beckstein.[11] The company celebrated its 10-year anniversary one year later. In the spring of 2010, voxeljet moved into a new administration building and production halls in Friedberg. During the same year, voxeljet was added to the list of Germany's top 100 innovators and received the "Top 100" seal of approval.[12]

73.1.4   Growth phase

In the year 2011, voxeljet introduced a series of technical innovations. In April 2011, the company opened up a new dimension of generative production methods with the VX4000 3D printing system.[13] This system makes it possible to produce objects with a size of 4 x 2 x 1 meters at a build speed that is up to three times faster than earlier systems, while maintaining the same resolution. At the international trade fair for foundry technologies "GIFA", voxeljet introduced the world's first continuous 3D printer, the VXC800.[14] The development of this continuous 3D printing technology represented the manufacturing of molds and models without tools. This machine generation runs the process steps "building" and "unpacking" in parallel, without having to interrupt system operations. Therefore, this printing system represents an important step towards industrial series production on the basis of a generative production process. In the same year, voxeljet celebrated the global premiere of its 3D printer VX1000 at EuroMold, the trade fair for tooling and mold-making, design and product development in Frankfurt am Main.[15] By combining high performance and a large build space, this printing system was able to meet the growing requirements of industry. In addition to these system innovations, the company also presented the newly developed material system Polypor type C in the year 2011. It allows voxeljet to meet customer demand for pure white plastic models. Moreover, this material also satisfies higher requirements regarding the stability and surface properties of the models.[16] A year later in 2012, the first VX1000 printing system was sold to the British company Propshop (Model Makers) Ltd.. It was the fifth voxeljet system that was in use in the United Kingdom. The movie industry opened a completely new customer market for voxeljet.[17] In the year 2012, voxeljet introduced its smallest system, the VX200, to the market. This printing system uses the same method as the larger series, but is very compact and easy to operate.

73.1.5   IPO

In the year 2013, the company went ahead with its IPO on the New York Stock Exchange. To this end, voxeljet technology GmbH was converted into a stock corporation that now operates as voxeljet AG. On 17 October 2013, voxeljet placed 6.5 million ADS on the NYSE at an issue price of US$13.[4][18] Five American Depositary Receipt corresponded to one share.[19] The IPO enabled voxeljet AG to take in US$64.5 million after deduction of the price discount granted to the issuing banks and issue costs.[20] Six months after the IPO, voxeljet AG generated another US$41.1 million as part of a capital increase, after deduction of the price discount granted to the issuing banks and issue costs. In the process, the company issued another three million American Depositary Shares at the New York Stock Exchange at a unit price of US$15.[5][19]

73.1.6   Globalization

To add to its global network of sales partners voxeljet AG began to set up its own international locations in the year 2014. On 1 October 2014, voxeljet AG took over the British company Propshop (Model Makers) Ltd., thus establishing its first location outside Germany. With the takeover, Propshop became a wholly owned subsidiary of voxeljet AG. The company, which specializes in the film and entertainment industry, had already gained experience with voxeljet printing systems when it purchased the VX1000 in the year 2012.[5][21][22] In the same year, voxeljet AG founded a new company in the USA. In January 2015, voxeljet began to operate a service center for the on-demand production of molds and models in Canton (Michigan), with the goal of reaching a printing capacity similar to the capacity at the home location in Friedberg by the end of 2016.[5][23]

73.2   Technology and process

73.2.1   Process

The technology was first developed at the Massachusetts Institute of Technology in 1993. And is generally known as the "Powder bed and inkjet head 3D printing". As usual in the additive manufacturing processes, the part to be printed is built up from many thin cross sections of the 3D model. An inkjet print head inkjet moves across a bed of powder, simultaneously putting down a liquid binding material. After that a thin layer of powder is extended across the completed section and the process is repeated several times with each layer adhering to the last.

73.3 See also

- List of 3D printer manufacturers

73.4 References

[1] "Unternehmen: Vorstand". voxeljet AG. Retrieved 2015-03-16.

[2] "voxeljet: Investor Relations: SEC Filings: Form 20F: 27/03/2015" (PDF). voxeljet AG. Retrieved 2015-04-08.

[3] "Bilanz aus 16 Jahren Münchener Businessplan Wettbewerb". UNITED NEWS NETWORK GmbH. Retrieved 2015-03-19.

[4] "Voxeljet: Was der Hersteller von 3D-Druckern drauf hat". wallstreet:online AG. Retrieved 2015-03-26.

[5] "Unternehmen: Firmenhistorie". voxeljet AG. Retrieved 2015-03-19.

[6] "Emeriti A-Z: Prof. Dr.-Ing. Dr.-Ing E.h. Joachim Heinzl". Technische Universität München. Retrieved 2015-03-19.

[7] "Erfolgsbroschüre Bayernkapital" (PDF; 1,1 MB). Bayern Kapital GmbH. Retrieved 2015-03-19.

[8] "Wirtschaftsstaatssekretär Pschierer gratuliert zum Börsengang". JONGO Webagentur. Retrieved 2015-03-13.

[9] "Portfolio: voxeljet AG". Franz Industriebeteiligungen AG. Retrieved 2015-03-23.

[10] "Ausstattung: Additive Fertigungsverfahren: Voxeljet VX500". Universität Rostock. Retrieved 2015-03-23.

[11] "Auszeichnung für voxeljet beim Bayerischen Innovationspreis 2008". UNITED NEWS NETWORK GmbH. Retrieved 2015-03-23.

[12] "Gütesiegel 'Top 100': Sechs Unternehmen aus der Kunststoffbranche ausgezeichnet". New Media Publisher GmbH. Retrieved 2015-03-25.

[13] "3D-Druck: Großformatiges 3D-Drucksystem generiert Objekte wirtschaftlich". Vogel Business Media GmbH & Co. KG. Retrieved 2015-03-25.

[14] "3D-Druck: Kontinuierlich arbeitender 3D-Drucker für die Kleinserienproduktion". Vogel Business Media GmbH & Co. KG. Retrieved 2015-03-25.

[15] "voxeljet VX1000 3D Printer: Industrial Scale Sand Casting & Prototyping". ENGINEERING.com. Retrieved 2015-03-25.

[16] "Ford Motor setzt auf Polypor C: In strahlendem Weiß". Konradin-Verlag Robert Kohlhammer GmbH. Retrieved 2015-03-25.

[17] "News: Filmindustrie setzt auf 3D-Druck". voxeljet AG. Retrieved 2015-03-25.

[18] "Update: voxeljet AG gibt Ausgabepreis für Börsengang bekannt". Dannes Solutions GmbH. Retrieved 2015-03-26.

[19] "Voxeljet: 3D-Drucker-Hersteller mit neuen Aktien". Boersengefluester.de. Retrieved 2015-03-26.

[20] "Furioses Börsendebüt: Voxeljet stürmt an die Börse". BÖRSENMEDIEN AG. Retrieved 2015-04-08.

[21] "3D-Druck: voxeljet AG übernimmt Propshop". vmm wirtschaftsverlag gmbh & co. kg. Retrieved 2015-04-08.

[22] "On the 1st October 2014 Propshop became a fully owned subsidiary of voxeljet". voxeljet UK Ltd. Retrieved 2015-04-08.

[23] "Voxeljet eröffnet Standort in den USA". Vogel Business Media GmbH & Co. KG. Retrieved 2015-04-08.

Chapter 74

Youmagine

YouMagine is an online repository of open source hardware designs that can be fabricated with a 3-D printer.[1]

74.1 References

[1] J. Biggs. YouMagine Brings Some Heat To The Free 3D Model Space TechCrunch Sept 17, 2013.

74.2 External links

- Official website

Chapter 75

Book:3D printing

75.1 3D printing

75.1.1 An overview

Overview 3D printing

3D bioprinting

3D tools and techniques 3D modeling

3D scanner

List of common 3D test models

Manufacturing processes Contour crafting

Direct metal laser sintering

D-Shape

Electron beam freeform fabrication

Fused deposition modeling

Laminated object manufacturing

Laser engineered net shaping

Magnetic 3D Bioprinting

Powder bed and inkjet head 3D printing

Stereolithography

Selective heat sintering

Selective laser sintering

Selective laser melting

Volumetric printing

Applications 3D printed firearms

3D-printed spacecraft

Building printing

Critical making

Injection molding

Rapid prototyping

Molding

Organ-on-a-chip

Tissue engineering

Related Molecular assembler

75.2 Text and image sources, contributors, and licenses

75.2.1 Text

- **3D printing** *Source:* https://en.wikipedia.org/wiki/3D_printing?oldid=689452013 *Contributors:* Bryan Derksen, Paul A, DavidWBrooks, Ronz, Julesd, Milkfish, Glenn, Radiojon, Cameronc, Samsara, Bevo, Bearcat, Robbot, Kizor, Nurg, Yosri, Gidonb, Mervyn, Jeroen, Xanzzibar, Jordon Kalilich, Alan Liefting, DavidCary, Wolfkeeper, BenFrantzDale, Timpo, Orangemike, RapidAssistant, Khalid hassani, Slurslee, Chowbok, Beland, Piotrus, Discospinster, Rich Farmbrough, Vsmith, Gronky, Bender235, Tom, Tgeller, Stesmo, Giraffedata, VBGFscJUn3, Sam Korn, Alansohn, Arlosuave, Mduvekot, Daniel.inform, Bios~enwiki, DreamGuy, BRW, Amorymeltzer, DV8 2XL, Drbreznjev, Dan100, Ceyockey, Tripodics, LeonWhite, Erich666, Pol098, Dennismk, Twthmoses, GregorB, Waldir, Dovid, Graham87, BD2412, Sjö, Rjwilmsi, Koavf, Hulagutten, Arabani, Bruce1ee, Graibeard, Lotu, DirkvdM, FlaBot, Ahasuerus, Gurch, BjKa, Kolbasz, Enon, Tedder, Kri, Ahunt, DVdm, Agamemnon2, UkPaolo, Bill Hewitt, Wester, Huw Powell, RussBot, Arado, Hede2000, GameFreak7744, Jaymax, Gaius Cornelius, Rsrikanth05, ALoopingIcon, Aeusoes1, Dahveed323, Natkeeran, Dbfirs, Kkmurray, Tonywalton, TransUtopian, Zzuuzz, Arthur Rubin, BorgQueen, VeryWetPaint, Mais oui!, DoriSmith, Jack Upland, ViperSnake151, JDspeeder1, Mardus, Samwilson, Luk, Palapa, Treesmill, SmackBot, Marc Lacoste, McGeddon, Jurriaan van Hengel, KVDP, Kintetsubuffalo, Yamaguchi⬚⬚, Gilliam, Ohnoitsjamie, FarMcKon, JMiall, Chris the speller, Jjalexand, Thumperward, Snori, Guypersonson, PrimeHunter, Mdwh, Victorgrigas, PureRED, Oni Ookami Alfador, Baa, Rcbutcher, Can't sleep, clown will eat me, Frap, Rrburke, GVnayR, Nakon, Ne0Freedom, Ian01, Marc-André Aßbrock, Salamurai, A5b, Arielco, Deepred6502, Byelf2007, SashatoBot, Ser Amantio di Nicolao, JzG, JorisvS, Minna Sora no Shita, Mgiganteus1, SpyMagician, Spiel, S zillayali, Noah Salzman, Freederick, Sorein~enwiki, DouglasCalvert, Wizard191, Iridescent, Aperium, Paul venter, Joseph Solis in Australia, Vanisaac, Eastlaw, FatalError, Joostvandeputte~enwiki, CRGreathouse, Tanthalas39, BKalesti, N2e, GargoyleMT, Twohlers, Maguffinator, Jordan Brown, Njlowrie, Cydebot, Sulka~enwiki, Mato, HokieRNB, Frzl, Gogo Dodo, Hebrides, The snare, Clovis Sangrail, Shirulashem, Plaasjaapie, Gaijin42, Yukichigai, Headbomb, Marek69, Nslsmith, Eljamoquio, Dawkeye, Nick Number, Gioto, Guy Macon, AnemoneProjectors, ABeatty, Masonba2000, Bakabaka, Danger, Alphachimpbot, Once in a Blue Moon, Lfstevens, Quarague, Ninahale, Ingolfson, JAnDbot, Barek, Awilley, Henkk78, TAnthony, Y2kcrazyjoker4, SiobhanHansa, Z22, Magioladitis, .snoopy., JNW, JamesBWatson, Oskay, Nikevich, Tonyfaull, Josiahseale, Ben Ram, Kawaputra, Enquire, Edward321, Oicumayberight, ElliAwesome, Philippe.beaudoin, Keith D, Jack007, Bus stop, R'n'B, CommonsDelinker, Verdatum, Tgeairn, Slash, Dbiel, NerdyNSK, Murmurr, Ellisbjohns, FrummerThanThou, Thatotherperson, Hodlipson, Crakkpot, Fountains of Bryn Mawr, Fklatt, Mufka, Sbierwagen, KylieTastic, Mkmori, STBotD, Rapidlaser, Bonadea, Squids and Chips, VolkovBot, Jeff G., JohnBlackburne, Jameslwoodward, Rubyuser, Butkiewiczm, Jay-so~enwiki, Philip Trueman, Oshwah, Mercurywoodrose, Kww, Tumblingsky, Crohnie, LeaveSleaves, DoktorDec, Raryel, Andy Dingley, PieterDeBruijn, Falcon8765, Nave.notnilc, Turgan, Vchimpanzee, TheBendster, RandallH, Austriacus, Lightbreather, Ezrado, SieBot, NHRef, Swliv, Paradoctor, VVVBot, Gerakibot, Claus Ableiter, Parhamr, Acasson, Chemako0606, Cwkmail, FunkMonk, Flyer22 Reborn, Nopetro, Aruton, Master munchies, Lightmouse, Redmercury82, StaticGull, Bcn0209, TrGordon, Denisarona, Escape Orbit, Martarius, Sfan00 IMG, KJG2007, ClueBot, Nick Churchill, Traveler100, Inition, Zeptomoon, Cab.jones, The Thing That Should Not Be, Mattgirling, Unbuttered Parsnip, Tomas e, Marzmich, Tjfr, Excirial, Threequarter-ten, Shahab.fm, BobKawanaka, Arjayay, Leecottrell, Another Believer, Michael751, Scalhotrod, DumZiBoT, Neuralwarp, XLinkBot, Jytdog, Scjules, Dthomsen8, WikHead, Sorathiya, AnneWiki, Wouterwolf~enwiki, Some jerk on the Internet, Non-dropframe, Friginator, Zellfaze, Grandscribe, Ronhjones, Jncraton, Aboctok, CanadianLinuxUser, MrOllie, Download, Lehid, SomeUsr, Techimo, Roux, Favonian, Nanzilla, Zorrobot, Jarble, Ettrig, Margin1522, Legobot, Korbnep, Luckas-bot, Yobot, WikiDan61, Themfromspace, Fraggle81, Reenem, FeydHuxtable, AnomieBOT, Momoricks, Efa, Jim1138, Jo3sampl, Flewis, Materialscientist, Asarkof, Citation bot, Techdoctor, LilHelpa, Sudoaptitude, Erud, Melmann, Nrpf22pr, Sunwin1960, Tallguy1982, TractusVicis, Crzer07, Almish80, Solphusion~enwiki, Softwarejonas, Riptide360, Microfilm, Brunonar, Alainr345, Shadowjams, Motsjo, A.amitkumar, GliderMaven, FrescoBot, Djeexpert, 3dcreationlab, DivineAlpha, Cannolis, Patafisik, DrilBot, Pinethicket, I dream of horses, Pj.vandendriessche, Tcarstensen, Jonesey95, NinjaCross, Tom.Reding, Lteschler, MastiBot, SpaceFlight89, 00zion00, Jandalhandler, Juanr2099, Skippy84, Trappist the monk, Xucy, Clarkcj12, Suburb 77, Reach Out to the Truth, Civic Cat, Onel5969, Mean as custard, RjwilmsiBot, Misconceptions2, Nmillerche, Sebastien Bailard, Steve03Mills, Emaus-Bot, Ariusturk, Eu6, Lucien504, Kronberger4, Cygnus1899, Dewritech, Peaceray, Klbrain, Solarra, Jmencisom, Winner 42, Baby Mama 2008, Werieth, Illegitimate Barrister, Josve05a, Brykl, EdMcCorduck, Davidhere40, MoireL.5522, Ὁ οἶστρος, Holdendesign, Briangarret, Iammak, Alphonse2, LastDodo, AManWithNoPlan, Tolly4bolly, Dranod, Brandmeister, Gsarwa, Kippelboy, Johnspencer, Cadjockey, Ego White Tray, Rangoon11, Wakebrdkid, Teapeat, Marketing2bot, Autodidact1, Keavon, ClueBot NG, Rich Smith, Accelerometer, Peter James, Gareth Griffith-Jones, Incompetence, Matthiaspaul, P.croaker, Greatrate, Manueldrama, Psubhashish, Widr, Heyandy889, Knives182, Anupmehra, Parthdu, CasualVisitor, Sameenahmedkhan, Glenjiman, Rapatan, IBrow1000, Helpful Pixie Bot, HMSSolent, Strike Eagle, Calabe1992, DBigXray, Technical 13, Jessica.yau, BG19bot, Beckyc24, Mohamed CJ, Fi63321, 2botmodelmaker, Juro2351, Mr.TAMER.Shlash, Nospildoh, ElphiBot, Frze, Jonathan Mauer, Zipzip50, Canoe1967, Phaneza, Praefulgidus, Sparthorse, DPL bot, John2bob, Bigmanbiggerman, Superfatcatgriz, Iliahs, Ginger Maine Coon, Fotoriety, 2botmarketing, Nanobliss, ThirthtonThithtertinton, Johnoly99, Makergear, AeroAlonso, Josephwoh, Usearch, Lindalise, Carliitaeliza, BattyBot, 3dfuture, Citing, Guanaco55, Autodidaktos, StarryGrandma, CelticWarrior49, MahdiBot, Cjripper, Cyberbot II, The Illusive Man, Jmcneil747, ChrisGualtieri, Embrittled, JNevil, EuroCarGT, Jos.scheepers, Felixphew, Paulaoceans, Serveradar2, IjonTichyIjonTichy, Uuu201, ExOne3D, Deezmaker, EagerToddler39, Kmm25, Blacksnark, Dexbot, Cervanza, Crc2012, Hmainsbot1, Mogism, 15mehr, Airwolf3d, Cerabot~enwiki, CLEChick, TwoTwoHello, Purplematty, Lugia2453, SFK2, Graphium, Tropicanamarie, CarolineKaup, MeliesArt, Janpih~enwiki, Eventorbot, Luli17, Ralphvb, Sam cfd, RandomLittleHelper, Signalbox, Hussulo, Joeinwiki, BlueRoll18, Austinn26, RaulyPatel01, Gaspardbos, Theo's Little Bot, Wjmcneil747, Troutmagnificent, Janus Savimbi, JulieAsarkofReece, Inntellektt, Dogenx, François Robere, Needle Mush, Andrewmtravels, Chibigold, Peabodybore, Tentinator, Everymorning, Amykam32, CosmosSoup, Rebstei, Rosenblumb1, Mskramer, EvergreenFir, Paul Whittaker Inovar, Bonmarly13, Backendgaming, Awartski, Sukumaar mane, SpecialGuy, Suswaltz, Kyle.maddox10, Branda.quintana, Alyssacles, Apeman2, Batboys, Comp.arch, Corinnecory, Koza1983, Thevideodrome, Ugog Nizdast, Soxtherobot, Tehben1, VelocityRap, Geekgirl72, Stephendavion, Mandruss, Ginsuloft, Manospeed, Rocco49228, Rcrumpf, Shashanksays123, RomyBallieux, RAF910, Acalycine, Cimorthing, MaloneyTim, M brinklow, Tyler-Greenberg, JohnAlexanderStewart, Noyster, DarkestElephant, Wilro, Nutterbutter54, UY Scuti, Stamptrader, YodasSpecies, Atanasov anton, Mauricio.delgado, Shellytel, Matsci2, Éffièdaligrh, ScienceFanatic100, Epic Failure, Itsalleasy, Wynnm5, Sliverpool9, Ralph80, Hy-Davo, Mindblaster6, TheEpTic, Suli92, Jeromic, Norseman08, Tonystarkman, HackerTon, NastyMan99, Lagoset, Robo3dprinter, JBCVS, Filedelinkerbot, Jrrfunding, SustainabilityAndy, Rory Top, Fyddlestix, Robinluniya, Ke48273, The Original Filfi, SpanglishArmado, Biggermig, Stakall, Xinai520, Lor, Verdana Bold, JezGrove, StephaniePBorger, Active 3D, Info202final, Astevens9, 3dgeek, Smfrayne, KH-1, Mario Castelán Castro, Ununuhuh, Merad17, ChamithN, Wolftribe, Nataraj.e, Mehari79, Alexliow, Cw585, Lgao33, Jvnap426, Kitty Hazel, Cellogoodbye, Pgold009, Kisg24, Amustaf2, JimmyWagger, Seamusprs, Highty1, Blistro, Sarr Cat, UnknownHenson, Mike Shostak, Gurragb, Jalbo01, Evanslyne, Wins.rajan, Travis836, Puramix, Zortwort, Claudio.cantone, CV9933, Truasami, Nickschwing, Adamiszczu, Krubinstein325, Kourousis, Fdm11, Infinite0694, Bananaforreal, Cassie meyer, Ivorycoasty, Empowering you, Gabrishl, GerraldGoogle, Nemesis2473, Macon11268865799?!', YesPretense, Alistairgray42, He3dwendy, ChemWarfare, ITGURU2015, Mackalna, Wikiowl66, TheEditor867, Peter3dsmith, Tripl5.creative, Mebakassahun, ProprioMe OW, Bijayabikramsamal, Eric2718, Researchmoz.us, Berting Li, Tlterp, Kaliardor, OrganicEarth, Orbit4447, JayLoerns, Ahouston5, Pos333, ChrisGuinvlx268, Shyannawalls, Msarraci, Lol master 6969, Casales1, Immcim2c, Blakenyguen, Suaveuser, Wdornenburg, Kristenhinzee, Jaxcab, Gui le chat and Anonymous: 746

- **3D publishing** *Source:* https://en.wikipedia.org/wiki/3D_publishing?oldid=685888132 *Contributors:* Ivan007, Ser Amantio di Nicolao, Jdaloner, Pjhermans, BG19bot, Mr mr ben, Mehari79, Nickschwing and Anonymous: 3

- **2BOT Physical Modeling Technologies** *Source:* https://en.wikipedia.org/wiki/2BOT_Physical_Modeling_Technologies?oldid=610990382 *Contributors:* Bearcat, Micru, Jpgordon, RussBot, Cydebot, Katharineamy, Don4of4, Wilhelmina Will, ImageRemovalBot, Eeekster, DASH-Bot, Widr, Beckyc24, 2botmodelmaker, Geilrules, PaintedCarpet, Kumioko and Anonymous: 1

- **3D bioprinting** *Source:* https://en.wikipedia.org/wiki/3D_bioprinting?oldid=678231235 *Contributors:* Bearcat, Ceyockey, Pol098, BD2412, Malcolma, Lenoxus, Fountains of Bryn Mawr, XLinkBot, Yobot, Anypodetos, AnomieBOT, Silviakuna, Jenks24, BG19bot, 220 of Borg, Dexbot, Wuerzele, OccultZone, Casvdschee, Zehranasser, SpanglishArmado, Kenzie flick, 3dprint, Teach380, Cassie meyer, Juanfran 44 and Anonymous: 15

- **3D Hubs** *Source:* https://en.wikipedia.org/wiki/3D_Hubs?oldid=680140397 *Contributors:* JHCaufield, Amatulic, KylieTastic, Niceguyedc, Yobot, Atilev, GermanJoe, BG19bot, Filedelinkerbot, Kisg24, Gfisherwils and Anonymous: 2

- **3D Manufacturing Format** *Source:* https://en.wikipedia.org/wiki/3D_Manufacturing_Format?oldid=686648124 *Contributors:* Bearcat, David-Cary, Rpyle731, Erich666, Viral sachde, 🔲🔲, Wgolf, BG19bot, JulianEP, Danyc0 and Pluto NP

- **3D Print Canal House** *Source:* https://en.wikipedia.org/wiki/3D_Print_Canal_House?oldid=684237951 *Contributors:* Smalljim, Bgwhite, Future Perfect at Sunrise, The Anomebot2, Wikimandia, CorenSearchBot, Boleyn, Yobot, AnomieBOT, Tufor, Editør, BG19bot, Crow, Ill Victims, Marthe2201 and Anonymous: 3

- **3D printed firearms** *Source:* https://en.wikipedia.org/wiki/3D_printed_firearms?oldid=679253310 *Contributors:* Tothebarricades.tk, Rich Farmbrough, Anastrophe, Gilliam, Derek R Bullamore, OnBeyondZebrax, N2e, Gaijin42, Headbomb, Rwessel, Andy Dingley, Meters, Lightbreather, XLinkBot, AnomieBOT, Miguel Escopeta, RjwilmsiBot, Josve05a, Petrb, ClueBot NG, Catlemur, BattyBot, HistoricMN44, Everymorning, Jonas Vinther, Rezin, RollaTroll, Amnichole and Anonymous: 9

- **3D printing marketplace** *Source:* https://en.wikipedia.org/wiki/3D_printing_marketplace?oldid=682290580 *Contributors:* Timpo, Ukexpat, Rich Farmbrough, Qwertyus, Bgwhite, Rwalker, Fram, Amatulic, Chris the speller, Dgw, Seaphoto, NatGertler, Andy Dingley, Bcn0209, Michaelspivey, Arjayay, Rankersbo, Pannini, AnomieBOT, Materialscientist, Citation bot, Mean as custard, Mhahnel, BG19bot, BattyBot, BurritoBazooka, AlexeevS, Jakec, Quenhitran, Atanasov anton, Mr mr ben, Pending, RPMarketplace, Kisg24, 3DLTlindsey, Picoline123, Bananaforreal, Jpmaterial, Dermotsul, JayLoerns and Anonymous: 22

- **3D-printed spacecraft** *Source:* https://en.wikipedia.org/wiki/3D-printed_spacecraft?oldid=665655139 *Contributors:* JorisvS, N2e, Jdaloner and BG19bot

- **Additive Manufacturing File Format** *Source:* https://en.wikipedia.org/wiki/Additive_Manufacturing_File_Format?oldid=664919081 *Contributors:* AnonMoos, Bearcat, Discospinster, Pmsyyz, Lothartklein, BD2412, Kri, Bgwhite, Smithkennedy, Cedar101, Viral sachde, Thumperward, Eastlaw, Bobcousins, Magioladitis, Avicennasis, Glrx, Katharineamy, Hodlipson, Zemoxian, Robenel, Galacticvoyager, JB Gnome, Yobot, WikiDan61, Wonderfl, AnomieBOT, FrescoBot, John of Reading, Gaxtrope, Frietjes, Nospildoh, Usearch, Devonak, Sjkelly, JohnAlexanderStewart, Atanasov anton and Anonymous: 12

- **Alumide** *Source:* https://en.wikipedia.org/wiki/Alumide?oldid=639359859 *Contributors:* Andrewman327, PamD, Eeekster, NintendoFan, ClueBot NG, Volcano dolphin, Captain Conundrum, Arthur.van.hoff and Anonymous: 2

- **Aluminum polymer composite** *Source:* https://en.wikipedia.org/wiki/Aluminum_polymer_composite?oldid=678506476 *Contributors:* Lfstevens, AnomieBOT, Jonesey95 and Dexbot

- **Arthur Mamou-Mani** *Source:* https://en.wikipedia.org/wiki/Arthur_Mamou-Mani?oldid=673649926 *Contributors:* TomStar81, Tony1, Dl2000, Hullaballoo Wolfowitz, Waacstats, Black Kite, Vrac, Yobot, Yngvadottir, J04n, BG19bot, Mamoumani, CookieMonster755, KratorOne and Anonymous: 8

- **Building printing** *Source:* https://en.wikipedia.org/wiki/Building_printing?oldid=680770222 *Contributors:* Khalid hassani, Vegaswikian, Natkeeran, Elkman, KVDP, Derek R Bullamore, N2e, Headbomb, Awilley, Yobot, AnomieBOT, Tom.Reding, K6ka, Ὁ οἶστρος, Wingman4l7, Ego White Tray, ClueBot NG, Virtualerian, IjonTichyIjonTichy, Lagoset, Monkbot, Ndijks, Trogluddite, Marthe2201 and Anonymous: 10

- **CandyFab** *Source:* https://en.wikipedia.org/wiki/CandyFab?oldid=583876460 *Contributors:* Milkfish, Mervyn, Db48x, Carcharoth, SmackBot, Kintetsubuffalo, Cydebot, Oskay, DumZiBoT, AnomieBOT, GliderMaven and Anonymous: 3

- **Cartesian coordinate robot** *Source:* https://en.wikipedia.org/wiki/Cartesian_coordinate_robot?oldid=686072950 *Contributors:* Andres, Nikola Smolenski, Charles Matthews, Wernher, ArnoldReinhold, Neko-chan, Drf5n, Spangineer, Japanese Searobin, Chochopk, Dysepsion, Robotics1, Anwar saadat, MichaelFrey, Swpb, CommonsDelinker, Jiuguang Wang, Andy Dingley, Steven Crossin, ClueBot, DBSpeakers, Addbot, Jewlake, MrOllie, DrilBot, HRoestBot, Chnsny, Cogiati, Midas02, ChuispastonBot, ClueBot NG, Widr, Danim, Atmabhola, Joeinwiki, Lagoset, KH-1 and Anonymous: 19

- **Complexity paradox** *Source:* https://en.wikipedia.org/wiki/Complexity_paradox?oldid=644441656 *Contributors:* Malcolma, Derek R Bullamore, Vanjagenije, SuperHamster, John of Reading, Aniijbod, Pontipietro and Anonymous: 1

- **Continuous Liquid Interface Production** *Source:* https://en.wikipedia.org/wiki/Continuous_Liquid_Interface_Production?oldid=683351230 *Contributors:* DrHow, Nikkimaria, Lfstevens, Shuvaev, Yobot, Brandmeister, HandsomeFella, BG19bot, Michael Barera, JakobSteenberg, AntonKjaer, Fuebaey and Anonymous: 2

- **D-Shape** *Source:* https://en.wikipedia.org/wiki/D-Shape?oldid=627446146 *Contributors:* Edward, Khalid hassani, GregorB, Rjwilmsi, Malcolma, SmackBot, N2e, Cydebot, Headbomb, Grundle2600, Blanchardb, Mdnahas, Tide rolls, AnomieBOT, Alvin Seville, GliderMaven, John of Reading, Ὁ οἶστρος, Ego White Tray, AlysonStoner5804, Virtualerian, ChrisGualtieri, Zapdos222, Monkbot and Anonymous: 1

- **Defense Distributed** *Source:* https://en.wikipedia.org/wiki/Defense_Distributed?oldid=685131278 *Contributors:* The Anome, Fred Bauder, Jpatokal, Bearcat, Wwoods, Pmsyyz, GregorB, Qwertyus, Czar, JHCaufield, Genjix, Arthur Rubin, Kamenev, Chris the speller, Autarch, Frap, Racklever, Mion, N2e, Cydebot, PamD, Gaijin42, Dream Focus, Magioladitis, Connor Behan, Keith D, CommonsDelinker, Andy Dingley, Lightbreather, Int21h, Cirt, XLinkBot, Dawynn, 0600Zulu, Tassedethe, Yobot, Sageo, AnomieBOT, Tucoxn, LilHelpa, RightCowLeftCoast, Vanished user kiij3irj4tihns, Wilton gorske, Miracle Pen, Canuckian89, Misconceptions2, GoingBatty, Your Lord and Master, Illegitimate Barrister, Tolly4bolly, Lead holder, BG19bot, WikiHannibal, Cliff12345, BattyBot, Mysterious Whisper, SoTotallyAwesome, Bitcoin, Jodosma, Nowheremano, Jackmcbarn and Anonymous: 39

- **DFM analysis for stereolithography** *Source:* https://en.wikipedia.org/wiki/DFM_analysis_for_stereolithography?oldid=685463719 *Contributors:* Edgar181, Chris the speller, Carter, Niceguyedc, AnomieBOT, I dream of horses, BG19bot and PranjalSingh IITDelhi

- **Digital materialization** *Source:* https://en.wikipedia.org/wiki/Digital_materialization?oldid=678287655 *Contributors:* Giraffedata, Onna, Cydebot, Falcon8765, Mild Bill Hiccup, Addbot, T0mt3, Aymankamelwiki, Lagoset and Anonymous: 1

- **Direct metal laser sintering** *Source:* https://en.wikipedia.org/wiki/Direct_metal_laser_sintering?oldid=685447643 *Contributors:* SimonP, Nurg, Khalid hassani, Snwright, SmackBot, Chris the speller, Snori, DMacks, Mgiganteus1, P199, Wizard191, N2e, Gaijin42, Hcobb, Guy Macon, Lfstevens, Fabrictramp, Lightmouse, Robenel, Three-quarter-ten, XLinkBot, AnomieBOT, Mikhael 666 mikhailovich, Jonesey95, Skippy84, Misconceptions2, GoingBatty, Ginger smiles, Plasticspro, Sachinvenga, Growmetal, Iitpkp, CasualVisitor, Themoother, Oafyman, Paul Whittaker Inovar, LCS check, AutoNoOpenMined, Spaghettimachine, Tstetler and Anonymous: 22

- **Distributed manufacturing** *Source:* https://en.wikipedia.org/wiki/Distributed_manufacturing?oldid=679284676 *Contributors:* Czar, LuckyLouie, Flyer22 Reborn, Duffbeerforme, Yobot, BG19bot, Mark viking, Batboys, Comp.arch, Fixture, Gigihit, AdamrobinsonCerasis, Richardbrt, Currystove3 and Anonymous: 7

- **Electron beam additive manufacturing** *Source:* https://en.wikipedia.org/wiki/Electron_beam_additive_manufacturing?oldid=685974916 *Contributors:* Heron, Ronz, Khalid hassani, Firsfron, CharlesC, Gaius Cornelius, DeadEyeArrow, SmackBot, Britiju, Kmarinas86, OrphanBot, Pwjb, Nagle, Wizard191, Twohlers, Cydebot, Alaibot, Mrbeardy, Parsecboy, MastCell, Dvmorris, AlphaEta, T. G. Savage, Flyingidiot, Anonymous Dissident, Andy Dingley, Fltnsplr, Aspects, Rapidmfg, Martinliljeberg, Three-quarter-ten, Thingg, Addbot, Yobot, Themfromspace, AnomieBOT, GrouchoBot, FrescoBot, ZéroBot, Mcmatter, Blue70s, ClueBot NG, BG19bot, Greenjackalope, Spaghettimachine, Selective laser melting and Anonymous: 31

- **Electron beam freeform fabrication** *Source:* https://en.wikipedia.org/wiki/Electron_beam_freeform_fabrication?oldid=685974521 *Contributors:* Brianhe, Tony1, Clayhalliwell, Kmarinas86, Nagle, Cydebot, Headbomb, Guy Macon, Ohms law, GimmeBot, Cnilep, Dabomb87, Alagrave, Spinoff, XLinkBot, Addbot, Mephiston999, Legobot, AnomieBOT, FrescoBot, Willardconnor, Jay1229, Dani7341, Danim, Canoe1967, Secondhand Work, François Robere, Suswaltz, CityEditors and Anonymous: 6

- **Fab@Home** *Source:* https://en.wikipedia.org/wiki/Fab%40Home?oldid=631163444 *Contributors:* Alan Liefting, Pol098, Kolbasz, Magioladitis, Hodlipson, LilHelpa, FrescoBot, Diannaa, Ymblanter, Shyncat, JeffreyiLipton and Anonymous: 8

- **Fab lab** *Source:* https://en.wikipedia.org/wiki/Fab_lab?oldid=688803055 *Contributors:* Ronz, Hectorthebat, Sj, Average Earthman, Spm, Khalid hassani, KevinBot, Nabla, Trevj, Brainy J, ABCD, DreamGuy, Schultz.Ryan, CharlesC, Radiant!, Jivecat, Quiddity, Oliverkeenan, Klonimus, Intgr, Bgwhite, Wavelength, Newbie222, Malcolma, SmackBot, Eskimbot, Gilliam, Bluebot, Thumperward, Olleolleolle~enwiki, Christefano, Ptroxler, Sander Säde, Joostvandeputte~enwiki, CmdrObot, GargoyleMT, Twelsht, Cydebot, Alaibot, Barticus88, Widefox, IndianGeneralist, Companje~enwiki, Albany NY, Gwern, Kasperh~enwiki, R'n'B, Andy Dingley, HairyWombat, Niceguyedc, Three-quarter-ten, Yohananw, XLinkBot, Addbot, Amys at mit, Zeldesse, Amonotron, Download, LaaknorBot, NittyG, Luckas-bot, WikiDan61, Themfromspace, KamikazeBot, AnomieBOT, Edobie, Quebec99, Xqbot, Foundrysmith, FrescoBot, Rboyer1, Jtlestum, T.Troublez, FVTCFablab, Pieter Cilliers, Harold horrible, RjwilmsiBot, EmausBot, ZéroBot, Crisp77, Johnsonj1, 汉武帝, GWIZsarasota, Wnc101496, FablabToulouse, ClueBot NG, Lokimee, Shaddim, Inventors Networked, Prototypeguru, Samueljon, Htcaballero, Manueldrama, Aghalim, KLBot2, ElphiBot, DPL bot, EdwardH, Usearch, Justincheng12345-bot, Igor.asonov, Makecat-bot, Joeinwiki, François Robere, Fixture, Lagoset, Elektrobody, Mrs Kartoshka, Yordhana, Juicy8287, Tannoz, Fablabncc and Anonymous: 114

- **Filaments evaluation protocol** *Source:* https://en.wikipedia.org/wiki/Filaments_evaluation_protocol?oldid=684093567 *Contributors:* Stesmo, Tony1, Yobot, GenQuest, NinjaCross, DPL bot, Kahtar, Filedelinkerbot and Anonymous: 2

- **Fused deposition modeling** *Source:* https://en.wikipedia.org/wiki/Fused_deposition_modeling?oldid=685106292 *Contributors:* Dreamyshade, Bryan Derksen, Ronz, Nurg, DavidCary, BenFrantzDale, RapidAssistant, Khalid hassani, Chowbok, Ehudshapira, KillerChihuahua, Bender235, Jaberwocky6669, Alga, John Vandenberg, Oarih, Halsteadk, Bhtooefr, BRW, Jim Mikulak, DV8 2XL, CharlesC, Rjwilmsi, Missmarple, Graibeard, Gurch, Raelx, Jzylstra, Whale~enwiki, Tony1, Terber, A Doon, SmackBot, Jurriaan van Hengel, LaurensvanLieshout, Edgar181, Thumperward, Tsca.bot, Blokkendoos, Quaeler, Zureks, N2e, GargoyleMT, Twohlers, Maguffinator, Cydebot, Matrix61312, Plaasjaapie, Headbomb, Guy Macon, Ninahale, Quickparts, .snoopy., Nikevich, Tgeairn, Valbyrne, VolkovBot, Fran Rogers, Andy Dingley, Mike the Mix, TheBendster, ImageRemovalBot, Slightlymighty, The Thing That Should Not Be, Journals88, Three-quarter-ten, Scalhotrod, Svachani, Addbot, Mplsap1970, Lightbot, ماني, Luckas-bot, Yobot, Themfromspace, AnomieBOT, Efa, Cubansmoothie, LilHelpa, Algspd, Karensams, Cleanstation-SRS, Grantmidnight, Stratocracy, Eugene-elgato, Some standardized rigour, Peterquale, GliderMaven, Gwideman, Mstrogoff, Jeffrd10, Macgeiger, Misconceptions2, Teapeat, OpticalBlimp, IBrow1000, Canoe1967, BattyBot, EuroCarGT, Rocknail, Ruby Murray, Dbrown9141, HyDavo, Clmthomas, Tbessler, Lagoset, Monkbot, Rory Top, WhatAboutThis0000, Mogie Bear, Sarr Cat, BrandoOk, ChemWarfare, Eric2718, Humbug26, Orbit4447 and Anonymous: 68

- **Fused filament fabrication** *Source:* https://en.wikipedia.org/wiki/Fused_filament_fabrication?oldid=685106296 *Contributors:* Michael Hardy, DavidCary, Ckielstra, Andy Dingley, Jdaloner, Niceguyedc, Priybrat, Yobot, Citation bot, Jenks24, BG19bot, Stamptrader, Lagoset, Kenhara, Wgn, Pgold009, BrandoOk and Anonymous: 4

- **IMakr** *Source:* https://en.wikipedia.org/wiki/IMakr?oldid=661999718 *Contributors:* Philg88, ImageRemovalBot, Yobot, BG19bot and Bananaforreal

- **Laminated object manufacturing** *Source:* https://en.wikipedia.org/wiki/Laminated_object_manufacturing?oldid=686975516 *Contributors:* BenFrantzDale, Zinnmann, Khalid hassani, Chowbok, CharlesC, Jurriaan van Hengel, LaurensvanLieshout, Betacommand, Sbmehta, Wizard191, Cydebot, .snoopy., VolkovBot, Addbot, GriffinDavid, Yobot, TaBOT-zerem, Univremonster, Materialscientist, Jesus Ultra, WikitanvirBot, Rapatan, Nanobliss, Epicgenius, Spaghettimachine and Anonymous: 11

- **Laser engineered net shaping** *Source:* https://en.wikipedia.org/wiki/Laser_engineered_net_shaping?oldid=677644851 *Contributors:* Michael Hardy, Delirium, Ehn, BenFrantzDale, Khalid hassani, Chowbok, PDH, Drajput, DV8 2XL, Srleffler, BirgitteSB, SmackBot, Jurriaan van Hengel, NCurse, Onceler, Pilotguy, Wizard191, Twohlers, Cydebot, Magioladitis, J Dezman, Ezrado, Addbot, Themfromspace, J04n, D'ohBot, Mean as custard, AutoNoOpenMined, Rory Top, Industrias Viwa and Anonymous: 8

- **Laser sintering of gold** *Source:* https://en.wikipedia.org/wiki/Laser_sintering_of_gold?oldid=661070419 *Contributors:* Khalid hassani, Vanjagenije, JL-Bot, ImageRemovalBot, Yobot, Ginger smiles, BattyBot, Jakec, Paqx and Anonymous: 1

- **List of 3D printer manufacturers** *Source:* https://en.wikipedia.org/wiki/List_of_3D_printer_manufacturers?oldid=689377015 *Contributors:* Iammaxus, Samsara, Ser Amantio di Nicolao, Postcard Cathy, Riptide360, Masum Ibn Musa, ArtZ72, Vogon Jelz, Michaeldarmani and Anonymous: 5

- **Local Motors** *Source:* https://en.wikipedia.org/wiki/Local_Motors?oldid=673757426 *Contributors:* McGeddon, Z22, Flyer22 Reborn, Yobot, AnomieBOT, Degen Earthfast, Seqqis, Lagoset, Tjbm, Unician, Richa b, Sergi Vidal Torrella and Anonymous: 3

- **Lyman filament extruder** *Source:* https://en.wikipedia.org/wiki/Lyman_filament_extruder?oldid=619249875 *Contributors:* Kolbasz, Rifleman 82, Squids and Chips, Andy Dingley, Dawynn, BattyBot, Batboys and Anonymous: 1

- **Made In Space, Inc.** *Source:* https://en.wikipedia.org/wiki/Made_In_Space%2C_Inc.?oldid=668826469 *Contributors:* Mercurywoodrose, Barleybob, Faolin42, BG19bot, KimberlyWylie, EoRdE6, 3D4space and Anonymous: 1

- **Magnetically assisted slip casting** *Source:* https://en.wikipedia.org/wiki/Magnetically_assisted_slip_casting?oldid=683351457 *Contributors:* Lfstevens

- **MatterHackers** *Source:* https://en.wikipedia.org/wiki/MatterHackers?oldid=611191709 *Contributors:* Ground Zero, BKalesti, JustAGal, Reddogsix and 3dgeek

- **MyMiniFactory** *Source:* https://en.wikipedia.org/wiki/MyMiniFactory?oldid=649634492 *Contributors:* Niceguyedc, Staszek Lem, Lakun.patra, Bananaforreal and Anonymous: 1

- **Nanophotonic coherent imager** *Source:* https://en.wikipedia.org/wiki/Nanophotonic_coherent_imager?oldid=663760169 *Contributors:* Rpyle731, Lfstevens, SporkBot and Wgolf

- **NovoGen** *Source:* https://en.wikipedia.org/wiki/NovoGen?oldid=648705068 *Contributors:* Ceyockey, Hiberniantears, Cydebot, Mr. Stradivarius, Yobot, FrescoBot and Anonymous: 2

- **Objet Geometries** *Source:* https://en.wikipedia.org/wiki/Objet_Geometries?oldid=617300071 *Contributors:* ELApro, Voxadam, SteveBaker, RadioFan, 2over0, SmackBot, Thumperward, Hmbr, Waacstats, Skier Dude, Quercus solaris, Guyonthesubway, Yobot, WikiDan61, AnomieBOT, Techdoctor, Armbrust, 00zion00, Mr.moyal, Mikola-Lysenko, Cloudscout, Snotbot, Helpful Pixie Bot, Usearch, Nimetapoeg, Savvy business and Anonymous: 5

- **Pinshape** *Source:* https://en.wikipedia.org/wiki/Pinshape?oldid=680842257 *Contributors:* Dthomsen8, Yobot, AnomieBOT, BattyBot, Iamsomswesome and Anonymous: 3

- **PLY (file format)** *Source:* https://en.wikipedia.org/wiki/PLY_(file_format)?oldid=688227919 *Contributors:* Pnm, Ldo, Saforrest, BenFrantzDale, Rpyle731, Bender235, Rjwilmsi, Chyel, SteveBaker, Wavelength, ALoopingIcon, EAderhold, SchfiftyThree, GargoyleMT, Cydebot, PhiLho, Tedickey, Hodlipson, JustinHagstrom, DumZiBoT, T68492, Addbot, OrlinKolev, AnomieBOT, Julini, Sirleto, Prosa100, Nospildoh, Dgirardeau, Furqanfurkan, Klosteraner and Anonymous: 12

- **Polyphenylsulfone** *Source:* https://en.wikipedia.org/wiki/Polyphenylsulfone?oldid=656239423 *Contributors:* Chowbok, Scottperry, A2Kafir, GregorB, DanMS, SmackBot, Edgar181, Cydebot, Kupirijo, Steevven1, FluffyWhiteCat, Yobot, Goatseeboy and Anonymous: 4

- **Powder bed and inkjet head 3D printing** *Source:* https://en.wikipedia.org/wiki/Powder_bed_and_inkjet_head_3D_printing?oldid=665267814 *Contributors:* Edward, Bearcat, Khalid hassani, SoWhy, Malcolma, Tony1, Argento, Cydebot, Headbomb, Mkmori, Bonadea, Cnilep, Tassedethe, Yobot, 4ndyD, John of Reading, Ego White Tray, Gavin.perch, Helpful Pixie Bot, Gorthian, Comp.arch, Monkbot, Spaghettimachine, KH-1, KarenHillwood89 and Anonymous: 4

- **Print the Legend** *Source:* https://en.wikipedia.org/wiki/Print_the_Legend?oldid=678715238 *Contributors:* Bill shannon, Back ache, Bovineboy2008, Americanfreedom, Yobot, 180beachview, Memeju, PD428 and Anonymous: 3

- **Projection micro-stereolithography** *Source:* https://en.wikipedia.org/wiki/Projection_micro-stereolithography?oldid=657130170 *Contributors:* Lfstevens

- **Proto BuildBar** *Source:* https://en.wikipedia.org/wiki/Proto_BuildBar?oldid=649696706 *Contributors:* ChuKat600 and Tangledupinbleu chs

- **Rapid prototyping** *Source:* https://en.wikipedia.org/wiki/Rapid_prototyping?oldid=689254297 *Contributors:* AxelBoldt, Deb, Rsabbatini, Michael Hardy, Skysmith, Goatasaur, Ahoerstemeier, Ronz, Extro, CatherineMunro, Pratyeka, Ehn, Vroman, Mydogategodshat, Timwi, Itai, Dpbsmith, Robbot, Tualha, Ancheta Wis, Psb777, Wolfkeeper, BenFrantzDale, Orangemike, Ssd, Zinnmann, Gracefool, Rdsmith4, Ehudshapira, Sonett72, Brianhe, Rama, BrokenSegue, Viriditas, DCEdwards1966, Pearle, Melaen, Danhash, Staeiou, Jim Mikulak, DV8

2XL, Oleg Alexandrov, Woohookitty, RHaworth, Cheesdude, Pol098, Knuckles, CharlesC, Btyner, Graham87, BD2412, Nightscream, Vegaswikian, Haya shiloh, Graibeard, Lotu, Ian Pitchford, Gurch, DVdm, YurikBot, Wavelength, Bhny, ALoopingIcon, Goffrie, Ndavies2, Scope creep, Zzuuzz, Nelson50, Teryx, Euke, Luk, SmackBot, Jurriaan van Hengel, KocjoBot~enwiki, Britiju, Gilliam, Ohnoitsjamie, Anwar saadat, Qwasty, Thumperward, Warpling, Guyjohnston, Freddyballo, Andreareinhardt, Man pl, CyrilB, Dicklyon, TastyPoutine, Wizard191, Courcelles, Tawkerbot2, Joostvandeputte~enwiki, CmdrObot, Dycedarg, Zureks, GargoyleMT, Twohlers, Cydebot, BillWeiss, Pascal.Tesson, Plaasjaapie, Satori Son, Thijs!bot, Nslsmith, LachlanA, AntiVandalBot, Widefox, Guy Macon, Seaphoto, Zigo1232, Masonba2000, Smartse, Alphachimpbot, Leuko, Skomorokh, Txomin, Albany NY, GoodDamon, LittleOldMe, SiobhanHansa, Quickparts, .snoopy., VoABot II, MastCell, Dvmorris, Lchrzan, Trusilver, Sageofwisdom, Rlsheehan, Murmurr, Vamcc, FrummerThanThou, Hodlipson, El monty, Quack 688, KylieTastic, Emalone, VolkovBot, DSRH, Sammiek23, Jay-so~enwiki, Philip Trueman, Sweetpea2007, Manufacturing, Duncan A Wood, Inventis, Madhero88, Kuczora, Andy Dingley, BrownBot, TheBendster, SieBot, 4wajzkd02, Frogpussy, Zo86, Foxtrotman, JackTheo, Emesee, Maxx88~enwiki, Firefly4342, Arrk, ClueBot, Avenged Eightfold, The Thing That Should Not Be, Three-quarter-ten, Arjayay, Dekisugi, BOTarate, Chaosdruid, Michael751, Aitias, Cassedu, Svachani, DumZiBoT, Scjules, Dthomsen8, James.barkley, Crazysane, MaterialGeeza, Jackienaylor, MrOllie, RTG, Roux, Lightbot, مانی, Zorrobot, Jarble, Gadibareli, Dengzhifan, Luckas-bot, Yobot, WikiDan61, AnomieBOT, Rubinbot, Materialscientist, Asarkof, Aff123a, Techdoctor, Xqbot, Capricorn42, Karensams, Almish80, J04n, Omnipaedista, GliderMaven, Anaday, Kagnie2, Squid661, MondalorBot, Akkida, X2bf3, Jhuglen, EmausBot, Helium4, CaptRik, Ό οἶστρος, Jasonjonesjones, ClueBot NG, Danim, CasualVisitor, KLBot2, Jessica.yau, BG19bot, Dsajga, M0rphzone, Vinaymn87, Nospildoh, MusikAnimal, Compfreak7, Veob66MI, Fspiceland, Usearch, Autodidaktos, Khazar2, Amirthinker, Mogism, Matheus Faria, Lingob, JulieAsarkofReece, Paul Whittaker Inovar, Hamoudafg, VelocityRap, Rcrumpf, ArdenM29, Lagoset, Musa Raza, OrganicEarth and Anonymous: 198

- **Recyclebot** *Source:* https://en.wikipedia.org/wiki/Recyclebot?oldid=664343786 *Contributors:* Bearcat, Clement Cherlin, BD2412, Pseudomonas, Tony1, Jesse Viviano, Magioladitis, Biscuittin, Fuddle, Smeagol 17, OlYeller21, BG19bot, Postmahomeson, Richardbrt, Mehari79, Iwilsonp and Highty1

- **RepRap Project** *Source:* https://en.wikipedia.org/wiki/RepRap_Project?oldid=685576957 *Contributors:* Bryan Derksen, Leandrod, Ezra Wax, TaranRampersad, Glenn, Nikola Smolenski, Ehn, Timwi, Val42, VikOlliver, Kd4ttc, Brouhaha, Wolfkeeper, BenFrantzDale, Micru, Esrogs, Jeremykemp, Ehudshapira, Zhmort, Wikkrockiana, Rich Farmbrough, Stesmo, Drf5n, Spitzl, Kocio, Voxadam, Oleg Alexandrov, Dandv, Pol098, Firien, GregorB, CharlesC, Susten.biz, Rjwilmsi, Letsburn00, Sarg, ColinHogben, SteveBaker, Crazytales, Nowa, Welsh, Jpbowen, JoeBorn, Wknight94, Gamboz, Mhi, Lynbarn, Knowledgeum, SmackBot, PaulWay, Zazaban, Jurriaan van Hengel, Chris the speller, Thumperward, Snori, OrphanBot, Guyjohnston, AdultSwim, Hu12, DogFog, Drvanthorp, Lcamtuf, Daedalus969, Servant74, Ilikefood, Cydebot, Robertinventor, Plaasjaapie, Thijs!bot, Widefox, Guy Macon, GermanX, Gwern, Zenhaus, Kateshortforbob, Krolco, HEL, Fountains of Bryn Mawr, WardXmodem, Remi0o, VolkovBot, Alex rosenberg35, Woodsstock, Owlofcreamcheese, Andy Dingley, PieterDeBruijn, Narayaan, 3velvet3, Wikiborg2, F.j.gaze, Sfan00 IMG, Plastikspork, Niceguyedc, Three-quarter-ten, Carriearchdale, 7, Sb66613, XLinkBot, Jytdog, RealityDysfunction, Duffbeerforme, Mortense, Kevin E Hawkins, Tsunanet, Grandscribe, K. Corbitt, Download, JB Gnome, Lightbot, Smeagol 17, Jarble, Luckas-bot, Yobot, UltraMagnus, AnomieBOT, Efa, Hiihammuk, Pellinore1, Brian H Wilson, Xqbot, GrouchoBot, Riptide360, Brunonar, Conseils, Floparallel, GliderMaven, AK2AK2, Full-date unlinking bot, HyperCapitalist, Wbortz, Ondra.pelech, EmausBot, Iamchenzetian, Xamuel, H3llBot, Dranod, SBaker43, Mektez, Ncmazvin, CocuBot, Lyla1205, Oinasz, Frietjes, Danim, BG19bot, John Cummings, BattyBot, Laodah, ChrisGualtieri, Aotefe, Mogism, Cerabot~enwiki, Luli17, RPlasticPirate, Joeinwiki, Batboys, DarkestElephant, Gigihit, HackerTon, Lagoset, Stakall, Richardbrt, Moonbeamio, Highty1, Blistro, Monkey3D, Proof Pro, Gondi56, Ivorycoasty, He3dwendy, Tripl5.creative, Reoieg and Anonymous: 129

- **Robocasting** *Source:* https://en.wikipedia.org/wiki/Robocasting?oldid=678516452 *Contributors:* Khalid hassani, Rjwilmsi, Ezrado, ImageRemovalBot, De728631, Ronhjones, Yobot, Dexbot, Buster Hatfield and Claudio.cantone

- **Sanguino3 G-Code** *Source:* https://en.wikipedia.org/wiki/Sanguino3_G-Code?oldid=538683087 *Contributors:* Bearcat, Cydebot, Kateshortforbob, Emteeoh and Anonymous: 1

- **Sculpteo** *Source:* https://en.wikipedia.org/wiki/Sculpteo?oldid=681847602 *Contributors:* Zundark, Ukexpat, Grafen, Derek R Bullamore, Bilby, Andy Dingley, Claus Ableiter, Yobot, SwisterTwister, EmausBot, Alphonse2, Johnspencer, Elizabeth75004, BattyBot, GClooney1234, Hmainsbot1, Bsquier, Atanasov anton, Vincepatrick, Cmetgy, 3DLTlindsey and Anonymous: 5

- **Selective heat sintering** *Source:* https://en.wikipedia.org/wiki/Selective_heat_sintering?oldid=687997387 *Contributors:* Bearcat, Yamla, Mgiganteus1, Yobot, Xqbot, Tomásdearg92, Crosstemplejay, Faizan, Spaghettimachine, Rory Top and Anonymous: 2

- **Selective laser melting** *Source:* https://en.wikipedia.org/wiki/Selective_laser_melting?oldid=656446722 *Contributors:* Fred Bauder, Julesd, Bearcat, Khalid hassani, Snori, P199, Wizard191, Cydebot, Fountains of Bryn Mawr, Phil Bridger, Lightmouse, Rapidmfg, Three-quarter-ten, PixelBot, Arjayay, Addbot, Mortense, AnomieBOT, LilHelpa, FrescoBot, Misconceptions2, Donner60, Bk314159, Growmetal, Eg0u4092, O.Koslowski, Djblacky1, Mogism, Wuerzele, RomyBallieux, BHauron, Rory Top, KLBelgium, TM1927, Jpmaterial and Anonymous: 10

- **Selective laser sintering** *Source:* https://en.wikipedia.org/wiki/Selective_laser_sintering?oldid=688212136 *Contributors:* Bryan Derksen, XJaM, Ronz, Ehn, Jeffq, Nurg, DocWatson42, BenFrantzDale, Everyking, Khalid hassani, Gzornenplatz, Sam Hocevar, Addicted2Sanity, Adashiel, D6, Alansohn, Cdc, Jim Mikulak, Scm83x, Nightscream, Graibeard, DirkvdM, Srleffler, Meawoppl, Jzylstra, Gaius Cornelius, Ms2ger, Cyrus Grisham, Mlibby, SmackBot, Jurriaan van Hengel, LaurensvanLieshout, Thumperward, Snori, Sadads, Thief12, Mion, Mgiganteus1, Ehheh, P199, Wizard191, GargoyleMT, Twohlers, Cydebot, MrMacMan, Plaasjaapie, Thijs!bot, Guy Macon, Ninahale, Albany NY, Quickparts, .snoopy., Fountains of Bryn Mawr, Dorftrottel, VolkovBot, AlysTarr, TheBendster, Flyer22 Reborn, Slightlymighty, 7Piguine, Tjfr, Alexbot, Three-quarter-ten, Rimefrost, Svachani, Addbot, MaterialGeeza, FSIM, Themfromspace, AnomieBOT, Materialscientist, Johnkm77, I dream of horses, Calmer Waters, Zachareth, EmausBot, Hunterp46, ZéroBot, ClueBot NG, Nobletripe, Helpful Pixie Bot, Thea10, Paul Whittaker Inovar, Flynn Milligan, Mindblaster6, Vieque, Rory Top and Anonymous: 68

- **Solid Ground Curing** *Source:* https://en.wikipedia.org/wiki/Solid_Ground_Curing?oldid=659266694 *Contributors:* Bearcat, JL-Bot, Schreiber-Bike, Yobot, BattyBot, Zay Yar Myint and Anonymous: 4

- **Sprout (computer)** *Source:* https://en.wikipedia.org/wiki/Sprout_(computer)?oldid=679086002 *Contributors:* Cloudbound, Frap, Lasersharp, JohnInDC, Jim.henderson, Mercurywoodrose, Timtempleton, AManWithNoPlan, Dbshort, Tdelga03 and Anonymous: 3

- **Stereolithography** *Source:* https://en.wikipedia.org/wiki/Stereolithography?oldid=688604536 *Contributors:* Eloquence, Bryan Derksen, Ronz, BenFrantzDale, Khalid hassani, Alves~enwiki, Mike Rosoft, Pmsyyz, ArnoldReinhold, Mofochickamo, Alansohn, DV8 2XL, Jeff3000, Tabletop, Miroku Sanna, Seidenstud, Graibeard, Lotu, YurikBot, Jzylstra, DRosenbach, Jurriaan van Hengel, LaurensvanLieshout, Thumperward, Ado, Tsca.bot, Can't sleep, clown will eat me, Jklin, Hu12, Wizard191, JohnCD, GargoyleMT, Twohlers, Rmallins, Njlowrie, Cydebot, Kupirijo, JFreeman, Sochwa, Headbomb, Bill0756, Guy Macon, Ninahale, Gatemansgc, Quickparts, .snoopy., MastCell, Sarahj2107, David Eppstein, KPD~enwiki, Lchrzan, Trusilver, Rlsheehan, 4johnny, FrummerThanThou, Kovo138, KylieTastic, VolkovBot, LokiClock, Sweetpea2007, Anonymous Dissident, Joshwilf, TheBendster, Frogpussy, ImageRemovalBot, ClueBot, 7Piguine, GorillaWarfare, Tjfr, Grrlfox, Johnson25006, Three-quarter-ten, Rhododendrites, Iohannes Animosus, SchreiberBike, Svachani, DumZiBoT, XLinkBot, Scjules, Rüdiger Marmulla, Addbot, MaterialGeeza, MrOllie, Jacobcolt, Yobot, WikiDan61, Themfromspace, Laserproto, AnomieBOT, Jim1138, Unara, Materialscientist, Techdoctor, Stratocracy, Firozinasab, GliderMaven, FrescoBot, Fgcity, DrilBot, Tom.Reding, Mstrogoff, Callanecc, Diannaa, GodfriedToussaint, Hunterp46, Bilbo571, Rocketrod1960, Mjbmrbot, ClueBot NG, Anagogist, Hallzer73, Epizarroso, 3dsystems, Phaneza, Adlhancock, Usearch, Stephenpnock, Tty780, 图图图图图, Hellowikielf, Rory Top, Jonguam, Ksaosa, Greatedits1, Distransient and Anonymous: 104

- **Stereolithography (medicine)** *Source:* https://en.wikipedia.org/wiki/Stereolithography_(medicine)?oldid=618281158 *Contributors:* Rsabbatini, Jeff3000, Thumperward, Cydebot, Squids and Chips, Mild Bill Hiccup, Horia Ionescu, Addbot, Yobot, AnomieBOT, DrilBot, Will Beback Auto, Mogism, Rjbibb, Monkbot and Anonymous: 2

- **STL (file format)** *Source:* https://en.wikipedia.org/wiki/STL_(file_format)?oldid=684620623 *Contributors:* AxelBoldt, Michael Hardy, Pnm, Ldo, Saforrest, Brouhaha, Wolfkeeper, BenFrantzDale, Orangemike, Tjic, Pearle, Ylem, Erich666, Marudubshinki, SteveBaker, YurikBot, Wavelength, Huw Powell, RussBot, Gaius Cornelius, ALoopingIcon, Asnatu wiki, Hcho~enwiki, Arthur Rubin, SmackBot, Slashme, Jurriaan van Hengel, LaurensvanLieshout, Thumperward, Tamfang, Frap, Marc-André Aßbrock, SeanAhern, Mwtoews, CyrilB, Hayttom, Brad Halls, Brutzman, Servant74, GargoyleMT, Goatchurch, Cydebot, Thijs!bot, Electron9, Guy Macon, Lovibond, Joachim Michaelis, UncommonArtist, VMwiki, Giles Bathgate, Hiplibrarianship, Gwern, Conquerist, Emeraude, Glrx, Hodlipson, Reelrt, VolkovBot, JohnBlackburne, MusicScience, Andy Dingley, TypoBoy, Cheakamus, BOTarate, DumZiBoT, T68492, Addbot, Bodysurfinyon, GyroMagician, Lihaas, Yobot, Wojciech mula, Cmurphy42, Wonderfl, AnomieBOT, VanishedUser sdu9aya9fasdsopa, Tfinc, Sidsoza, Efa, LilHelpa, Xqbot, Lewix, Alan8, Materialise, Lochneil, GliderMaven, Meshing, Bpmcneilly, Jamesmwiki, DrilBot, ISCIX-Ex, Dinamik-bot, Limited Atonement, Hunterp46, Holdendesign, L Kensington, WaterCrane, Mikhail Ryazanov, ClueBot NG, FMax, Helpful Pixie Bot, Drsimonz, BG19bot, Neptune's Trident, Kvrantzaliev, Nospildoh, Habitmelon, Compfreak7, Mr.Anderson-Queens-AQ, ChrisGualtieri, YFdyh-bot, Esqueue, Dgirardeau, Sjkelly, RaoOfPhysics, JaconaFrere, Mr mr ben, Lagoset, Juxiliary, JayLoerns and Anonymous: 82

- **Strati (automobile)** *Source:* https://en.wikipedia.org/wiki/Strati_(automobile)?oldid=675234206 *Contributors:* Julesd, Kaldari, Koavf, Nikkimaria, Chris the speller, Z22, Victuallers, Yobot, Magic119, ClueBot NG, Widr, Northamerica1000, Andyhowlett, Epicgenius, Lagoset and Anonymous: 6

- **Thingiverse** *Source:* https://en.wikipedia.org/wiki/Thingiverse?oldid=663992876 *Contributors:* Zundark, Pnm, Timrollpickering, Micru, Ukexpat, Aplumb, RussBot, Nowa, Thumperward, Ser Amantio di Nicolao, Marcuscalabresus, Ingolfson, Steven Walling, Jehan60188, Andy Dingley, Claus Ableiter, Trivialist, Duffbeerforme, Addbot, Yobot, Unbitwise, RjwilmsiBot, Sheeana, Ό οἶστρος, UncleIze, Helpful Pixie Bot, Tart2000, Jochiat, Tentinator, Bsquier, Kolergy, Lagoset, KH-1, Cmetgy, 3DLTlindsey and Anonymous: 17

- **Threeding** *Source:* https://en.wikipedia.org/wiki/Threeding?oldid=653804075 *Contributors:* Bgwhite, Fram, Chris the speller, DGG, Rankersbo, GoingBatty, Atanasov anton, TheQ Editor and ArturZ72

- **TRI-D (rocket engine)** *Source:* https://en.wikipedia.org/wiki/TRI-D_(rocket_engine)?oldid=630846551 *Contributors:* Andy Dingley, Niceguyedc, Yobot, Jim1138, I dream of horses, Misconceptions2, Gomu gomu no pistol and Anonymous: 2

- **Ultrasonic consolidation** *Source:* https://en.wikipedia.org/wiki/Ultrasonic_consolidation?oldid=637755937 *Contributors:* Rich Farmbrough, Voxadam, Wizard191, Cydebot, .snoopy., Yobot, Mmrjf3, Paqx and Anonymous: 2

- **Voxeljet** *Source:* https://en.wikipedia.org/wiki/Voxeljet?oldid=685580734 *Contributors:* Chris the speller, Vchimpanzee, MenoBot, Niceguyedc, Yobot, Riptide360, EarwigBot, John of Reading, Jeraphine Gryphon, Sulfurboy, Samwalton9, FoCuSandLeArN, UY Scuti, Davidwikipedia94 and Anonymous: 3

- **Youmagine** *Source:* https://en.wikipedia.org/wiki/Youmagine?oldid=648893604 *Contributors:* Bearcat, Rpyle731, Wikiwarrior77 and Ivorycoasty

- **Book:3D printing** *Source:* https://en.wikipedia.org/wiki/Book%3A3D_printing?oldid=627430814 *Contributors:* Headbomb and Physistsheep

75.2.2 Images

- **File:2006_AEGold_Proof_Obv.png** *Source:* https://upload.wikimedia.org/wikipedia/commons/7/76/2006_AEGold_Proof_Obv.png *License:* Public domain *Contributors:* United States Mint *Original artist:* United States Mint

- **File:3-d_printed_flower_model.jpg** *Source:* https://upload.wikimedia.org/wikipedia/commons/4/4c/3-d_printed_flower_model.jpg *License:* CC BY-SA 4.0 *Contributors:* Own work *Original artist:* Jonathan Mauer

- **File:3D_Extruder_Driving_Force.png** *Source:* https://upload.wikimedia.org/wikipedia/commons/4/4f/3D_Extruder_Driving_Force.png *License:* CC BY-SA 4.0 *Contributors:* Own work *Original artist:* Priybrat

- **File:3D_Printer_Extruder.png** *Source:* https://upload.wikimedia.org/wikipedia/commons/e/e8/3D_Printer_Extruder.png *License:* CC BY-SA 3.0 *Contributors:* Microsoft Power Point *Original artist:* Priybrat

- **File:3D_printer2.jpg** *Source:* https://upload.wikimedia.org/wikipedia/commons/5/5f/3D_printer2.jpg *License:* CC BY-SA 3.0 *Contributors:* Own work *Original artist:* Tiia Monto

- **File:3dprinter.jpg** *Source:* https://upload.wikimedia.org/wikipedia/commons/d/d8/3dprinter.jpg *License:* CC-BY-SA-3.0 *Contributors:* Transferred from en.wikipedia to Commons by Wuzur using CommonsHelper. *Original artist:* Rsabbatini at English Wikipedia

- Nuvola_apps_emacs.png *Original artist:* Nuvola_apps_emacs.png: David Vignoni

- **File:Guardians_of_Time_sculpture_Manfred_Kielnhofer_3d_printing.JPG** *Source:* https://upload.wikimedia.org/wikipedia/commons/4/4e/Guardians_of_Time_sculpture_Manfred_Kielnhofer_3d_printing.JPG *License:* CC BY-SA 3.0 *Contributors:* Own work *Original artist:* Kronberger4

- **File:Hookem_hand.svg** *Source:* https://upload.wikimedia.org/wikipedia/commons/9/9c/Hookem_hand.svg *License:* CC-BY-SA-3.0 *Contributors:* Transferred from en.wikipedia to Commons. *Original artist:* The original uploader was Gustavb at English Wikipedia

- **File:Hp_9862a.jpg** *Source:* https://upload.wikimedia.org/wikipedia/commons/1/12/Hp_9862a.jpg *License:* CC BY-SA 4.0 *Contributors:* Own work *Original artist:* Florian Schäffer

- **File:HumanRightsLogo.svg** *Source:* https://upload.wikimedia.org/wikipedia/commons/d/d3/HumanRightsLogo.svg *License:* Copyrighted free use *Contributors:* http://humanrightslogo.net/ *Original artist:* Predrag Stakić, released by http://humanrightslogo.net/

- **File:Hyperboloid_Print.ogv** *Source:* https://upload.wikimedia.org/wikipedia/commons/5/5d/Hyperboloid_Print.ogv *License:* CC BY 3.0 *Contributors:* http://www.youtube.com/watch?v=1213kMys6e8 *Original artist:* Video: OhmEye. Object file: MaskedRetriever

- **File:IMakr_Logo.jpeg** *Source:* https://upload.wikimedia.org/wikipedia/en/7/7d/IMakr_Logo.jpeg *License:* CC-BY-3.0 *Contributors:* ? *Original artist:* ?

- **File:IMakr_store_London.jpeg** *Source:* https://upload.wikimedia.org/wikipedia/en/e/e0/IMakr_store_London.jpeg *License:* CC-BY-3.0 *Contributors:* ? *Original artist:* ?

- **File:I_robot_car.jpg** *Source:* https://upload.wikimedia.org/wikipedia/commons/a/ac/I_robot_car.jpg *License:* CC BY 2.0 *Contributors:* Flickr *Original artist:* Eirik Newth

- **File:Icon-gears2.svg** *Source:* https://upload.wikimedia.org/wikipedia/commons/0/06/Icon-gears2.svg *License:* Public domain *Contributors:* Image:Icon-gears.png *Original artist:* vector image: Gothika

- **File:Increase2.svg** *Source:* https://upload.wikimedia.org/wikipedia/commons/b/b0/Increase2.svg *License:* Public domain *Contributors:* Own work *Original artist:* Sarang

- **File:Laminated_object_manufacturing.png** *Source:* https://upload.wikimedia.org/wikipedia/commons/a/a1/Laminated_object_manufacturing.png *License:* CC BY-SA 3.0 *Contributors:* Own work *Original artist:* LaurensvanLieshout

- **File:LampFlowchart.svg** *Source:* https://upload.wikimedia.org/wikipedia/commons/9/91/LampFlowchart.svg *License:* CC-BY-SA-3.0 *Contributors:* vector version of Image:LampFlowchart.png *Original artist:* svg by Booyabazooka

- **File:Large_delta-style_3D_printer.jpg** *Source:* https://upload.wikimedia.org/wikipedia/commons/d/d0/Large_delta-style_3D_printer.jpg *License:* CC BY-SA 4.0 *Contributors:* Own work *Original artist:* Z22

- **File:Laser_sintered_bracelet_gold.jpeg** *Source:* https://upload.wikimedia.org/wikipedia/commons/a/a1/Laser_sintered_bracelet_gold.jpeg *License:* CC BY-SA 3.0 *Contributors:* Own work *Original artist:* Towe jewels

- **File:Lens_triplet.svg** *Source:* https://upload.wikimedia.org/wikipedia/commons/7/74/Lens_triplet.svg *License:* CC-BY-SA-3.0 *Contributors:* Image:Lens triplet.png *Original artist:* Panther, Antilived

- **File:Letter-from-Department-of-State-to-Defense-Distributed.pdf** *Source:* https://upload.wikimedia.org/wikipedia/commons/0/04/Letter-from-Department-pdf *License:* Public domain *Contributors:* betabeat *Original artist:* United States Department of State

- **File:Lyman-filament-extruder.jpg** *Source:* https://upload.wikimedia.org/wikipedia/commons/3/32/Lyman-filament-extruder.jpg *License:* CC BY 1.0 *Contributors:* http://www.thingiverse.com/thing:34653 *Original artist:* Hugh Lyman

- **File:MakerBot_ThingOMatic_Bre_Pettis.jpg** *Source:* https://upload.wikimedia.org/wikipedia/commons/7/75/MakerBot_ThingOMatic_Bre_Pettis.jpg *License:* CC BY 2.0 *Contributors:* Flickr, specific image page URL: http://www.flickr.com/photos/bre/3458247336/ *Original artist:* Bre Pettis

- **File:Maker_Faire_2008_San_Mateo_206.JPG** *Source:* https://upload.wikimedia.org/wikipedia/commons/b/b8/Maker_Faire_2008_San_Mateo_206.JPG *License:* CC BY 3.0 *Contributors:* Own work *Original artist:* ShakataGaNai

- **File:Merge-arrow.svg** *Source:* https://upload.wikimedia.org/wikipedia/commons/a/aa/Merge-arrow.svg *License:* Public domain *Contributors:* ? *Original artist:* ?

- **File:Mergefrom.svg** *Source:* https://upload.wikimedia.org/wikipedia/commons/0/0f/Mergefrom.svg *License:* Public domain *Contributors:* ? *Original artist:* ?

- **File:Microelectronics_stub.svg** *Source:* https://upload.wikimedia.org/wikipedia/commons/c/c3/Microelectronics_stub.svg *License:* LGPL *Contributors:* Integrated circuit icon.svg: <a href='//commons.wikimedia.org/wiki/File:Integrated_circuit_icon.svg' class='image'><img alt='Integrated circuit icon.svg' src='https://upload.wikimedia.org/wikipedia/commons/thumb/8/8c/Integrated_circuit_icon.svg/50px-Integrated_circuit_icon.svg.png' width='50' height='50' srcset='https://upload.wikimedia.org/wikipedia/commons/thumb/8/8c/Integrated_circuit_icon.svg/75px-Integrated_circuit_icon.svg.png 1.5x, https://upload.wikimedia.org/wikipedia/commons/thumb/8/8c/Integrated_circuit_icon.svg/100px-Integrated_circuit_icon.svg.png 2x' data-file-width='180' data-file-height='180' /></a> *Original artist:* Integrated_circuit_icon.svg: Everaldo Coelho and YellowIcon

- **File:Miniature_human_face_models_made_through_3D_Printing_(Rapid_Prototyping).jpg** *Source:* https://upload.wikimedia.org/wikipedia/commons/3/35/Miniature_human_face_models_made_through_3D_Printing_%28Rapid_Prototyping%29.jpg *License:* CC BY-SA 3.0 *Contributors:* Own work *Original artist:* S zillayali

- **File:NASA_EBF3_2007_test.jpg** *Source:* https://upload.wikimedia.org/wikipedia/commons/5/53/NASA_EBF3_2007_test.jpg *License:* Public domain *Contributors:* http://www.nasa.gov/centers/langley/news/researchernews/rn_zero-g5.18.06.html *Original artist:* NASA/Johnson Space Center

75.2.3 Content license